Design, Ecology, Politics

'*Design, Ecology, Politics* is a powerful contribution. It is a thoroughly innovative, provocative and confrontational approach to pressing twenty-first century challenges at the human–environment interface. Boehnert has produced a compelling book which expands our considerations of design and ecological literacy in the complex socio-economic systems where we find "home". I enthusiastically recommend this work to those interested in charting productive and sustainable pathways through today's ecological, social and cultural challenges.'

Maxwell Boykoff, **Associa sor of Environmental Studies at the University of Colorado-Boulder, USA**

'*Design, Ecology,* Politics commands design educators to ground their practice in critically engaged ecological literacy. Boehnert's deeply textured and carefully crafted clarion call should be read by all who design on our earth. And it is a must for current and future planners.'

Christopher Silver, **Professor of Urban and Regional Planning at the University of Florida, USA**

Design, Ecology, Politics

Towards the Ecocene

Joanna Boehnert

Bloomsbury Academic
An imprint of Bloomsbury Publishing Plc

B L O O M S B U R Y
LONDON · OXFORD · NEW YORK · NEW DELHI · SYDNEY

Bloomsbury Academic
An imprint of Bloomsbury Publishing Plc

50 Bedford Square	1385 Broadway
London	New York
WC1B 3DP	NY 10018
UK	USA

www.bloomsbury.com

BLOOMSBURY and the Diana logo are trademarks of Bloomsbury Publishing Plc

First published 2018

© Joanna Boehnert, 2018

Joanna Boehnert has asserted her right under the Copyright, Designs and Patents Act, 1988, to be identified as Author of this work.

All rights reserved. No part of this publication may be reproduced or transmitted in any form or by any means, electronic or mechanical, including photocopying, recording, or any information storage or retrieval system, without prior permission in writing from the publishers.

No responsibility for loss caused to any individual or organization acting on or refraining from action as a result of the material in this publication can be accepted by Bloomsbury or the author.

British Library Cataloguing-in-Publication Data
A catalogue record for this book is available from the British Library.

ISBN	HB:	978-1-4725-8861-6
	PB:	978-1-4725-8860-9
	ePDF:	978-1-4725-8863-0
	ePub:	978-1-4725-8862-3

Library of Congress Cataloging-in-Publication Data
Names: Boehnert, Joanna, author.
Title: Design, ecology, politics : towards the ecocene / Joanna Boehnert.
Description: New York : Bloomsbury Academic, An imprint of Bloomsbury Publishing Plc, 2018. | Includes bibliographical references.
Identifiers: LCCN 2017043791 | ISBN 9781472588616 (hardback)
Subjects: LCSH: Communication in design. | Design—Environmental aspects. | Design—Social aspects.
CLASSIFICATION: LCC NK1510 .B64 2018 | DDC 745.4—DC23 LC RECORD AVAILABLE AT HTTPS://LCCN.LOC.GOV/2017043791

Cover design: Clare Turner, Joanna Boehnert, Lazaros Kakoulidis and Tzortzis Rallis

Typeset by Integra Software Services Pvt. Ltd.
Printed and bound in India

Contents

List of Figures — vi
Acknowledgements — viii
Introduction: Within and Beyond Error — 1

Part 1 Design — 13

1 Design Theory 101 — 15
2 Design as Symbolic Violence — 27
3 Design versus The Design Industry — 38

Part 2 Ecology — 49

4 Ecological Theory 101 — 51
5 Epistemology Error — 62
6 Ecological Literacy — 74
7 Ecoliterate Design — 88
8 Ecological Movements — 102
9 Ecological Perception 1 – Theory — 113
10 Ecological Perception 2 – Practice — 121
11 Ecological Identity — 132

Part 3 Politics — 143

12 Social Marketing — 145
13 The Green Economy — 152
14 The Technofix — 160
15 Data/Knowledge Visualization — 170

Conclusion: Towards the Ecocene — 182
References — 186
Index — 201

List of Figures

I.1	*The Three Ecologies A.* EcoLabs. 2017.	3
I.2	*The Three Ecologies (B, C + D).* EcoLabs. 2017.	4
I.3	*Planetary Boundaries*, detail of *An Audit of Development*. EcoLabs + Tzortzis Rallis, Lazaros Kakoulidis, 2013.	6
I.4	*An Audit of Development*. EcoLabs, Tzortzis Rallis and Lazaros Kakoulidis, 2013.	8
I.5	*Earth System Trends*. The Trajectory of the Anthropocene: The Great Acceleration. Steffen et al. 2015.	9
1.1	*A Continuum of Design Approaches* (Irwin et al. 2015).	23
1.2	*Transition Design Framework* (Irwin et al. 2015).	24
2.1	*The Occupy Times,* #07. Lazaros Kakoulidis and Tzortzis Rallis, 2011.	34
2.2	*The Occupy Times,* #20. Lazaros Kakoulidis and Tzortzis Rallis, 2013.	35
2.3	*The Poor Man's Guardian. A Weekly Paper for the People.*	35
2.4	Shields (front and back) from the Climate Camp Heathrow occupation. Featured on the cover of *Creative Review* and in the V&A's *Disobedient Objects* exhibition. PSC Photography, 2007.	36
2.5	Climate Camp shields in use at the Heathrow Camp for Climate Action 2007. Kristian Buus, 2009.	37
3.1	*The Embedded Economy*. EcoLabs, 2015.	42
3.2	*The Stable/Unstable Constellations of the Three Domains*. EcoLabs, 2015 – following Shiva (2005, 52).	42
4.1	*Phaeodari. Kunstformen der Natur*, Plate 61, Ernest Haeckel, 1904.	54
6.1	*Three Parts of a Paradigm* (following Sterling 2003, 2011). Boehnert, 2017.	78
6.2	*Perception, Conception, Action*. EcoLabs, 2017.	80
6.3	*Design Interventions: The Perception to Practice Cycle*. EcoLabs, 2017.	81
7.1	*The Four Design Models for Circularity*. RSA.	90
7.2	*Tracing the Global Flow of Energy from Fuel to Service*.	91
7.3	Per capita consumption-based emissions, Meng et al. 2016.	94
7.4	Production-based emissions linked to foreign consumption, Meng et al. 2016.	96
7.5	*The Game Plan. My 2007 Diet.* Saul Griffins, 2008.	97
7.6	*The Steady State Economy: A Totem of Real Happiness*. EcoLabs and Angela Morelli, 2009.	99
10.1	Ways to graphically represent relations between two entities. Birger Sevaldson, 2013.	122
10.2	*Types of systemic relations*. Birger Sevaldson, 2016.	122

10.3	*Greenpeace oceans campaign. How's my fishing?* Density Design and Greenpeace UK.	124
10.4	*The Small-Scale Energy Harvesting project.* Designer: Adrian Paulsen, Advisor: Birger Sevaldson, 2009.	126
10.5	Flow of actions during an oil spill accident.	127
10.6	*The Syntax of a New Language.* Manuel Lima. 2011.	128
10.7	*Ecological footprint per country, per person,* 2008. WWF 2012.	129
11.1	*Ecological Self.* EcoLabs, 2017.	132
11.2	*Circumflex Model of Value Systems.* EcoLabs, 2017.	133
11.3	*The Psychology of Environmental Crisis: Denial Bubbles.* EcoLabs, 2017.	137
11.4	*The Psychology of Environmental Crisis: Trajectory of Denial.* EcoLabs, 2017.	138
11.5	*We Are Nature Protecting Itself.*	142
12.1	Hopenhagen advert: A Bottle of Hope. 2009.	148
12.2	Hopenhagen installation. Photos by Kristian Buus, 2009.	148
13.1	*Economic Approaches to the Environment.* EcoLabs, 2017.	154
13.2	*Conceptions of Human–Natural Relations: A Hierarchy of Systems.* EcoLabs, 2014.	155
13.3	*Costing the Earth* by Information is Beautiful Studio, 2011.	157
14.1	*Degrowth vs Ecomodernism.* EcoLabs, 2016.	165
15.1	Tweet and chart: *Lynchings in the United States, 1882–1969.* Max Roser, 2016.	173
15.2	*Natural Catastrophes.* Max Roser, 2016.	174
15.3	*Wisdom/Knowledge/Information/Data Triangle.* EcoLabs, 2017.	176
15.4	*Rise and Fall of Issues in UNFCCC Negotiations, 1995–2013.* Climaps by Emaps, 2014.	177
15.5	*Mapping Climate Communication, No. 1 The Climate Timeline.* Boehnert, 2014.	178
15.6	Sketch for *Network of Actors.* Boehnert, 2014.	179
15.7	*Mapping Climate Communication, No. 2 Network of Actors.* Boehnert, 2014.	180
C.1	*Another World is Possible.*	183

Acknowledgements

This book is the product of a decade's worth of work formulating a stronger ecological theory for design. It started as my doctoral research at the University of Brighton. I am grateful to the Arts and Humanities Research Council who provided the funding that made this initial work possible. I am indebted to my PhD supervisors Jonathan Chapman and Julie Doyle for help nurturing the ideas embedded in this book. Later, during a post-doc at the Cooperative Institute for Research in Environmental Sciences (CIRES) at the University of Colorado Boulder, Max Boykoff provided constructive encouragement at the book proposal stage. Back in the UK, I worked as a gardener at Tapeley Park with permaculture legend Jenny Hayns while writing the first draft. The final version was written during my time as a part-time Research Fellow at the Centre for Research and Education in Arts and Media (CREAM) at the University of Westminster. I must thank Tom Corby and Christian Fuchs for their intellectual support over this period. Many others have helped with ideas, critique, collaborations and help accessing funding over the years. Among them are Emma Dewberry, Guy Julier, Karin Jaschke, Isis Nunez Ferrera, Jon Goodbun, Doug Specht, Rosie Thomas, Bianca Elzenbaumer, Idil Gaziulusoy, Jonathan Crinion, Christopher Pierce, Donal O'Driscoll and Damian White. I would like to express my special appreciation to Michael Reinsborough who provided invaluable feedback, editing and suggestions on issues of the environment, science and technology. Although ideas in this book may have their origins from conversations with many people, any errors or shortcomings are my own.

I owe a great deal to social movements that have informed my understanding of environmental problems, social justice and socio-political change. Thanks to Paul Chandler who provided the solar and wind power at Climate Camp, taught me how to approach energy issues critically and arranged hiking holidays. Thanks to Noel Douglas, Tony Credland, Lazaros Kakoulidis, Tzortzis Rallis and Bianca Elzenbaumer for their work creating new forms of design activism and politicizing design theory and practice. I am also grateful to Lazaros and Tzortzis for their artwork that was developed into the cover of this book. Thanks to Angela Morelli for her artwork *The Steady State Economy: A Totem of Real Happiness* – and to all designers who contributed artwork for this book. I want to thank to everyone at Bloomsbury who helped put this book into production especially Manikandan Kuppan, Ken Bruce, Abbie Sharman, Claire Constable and Rebecca Burden.

My family and friends have provided indispensable encouragement and support. My father Gunnar taught me how to consider historical context and this outlook is integral to the perspective taken in this book. Thanks to those who graciously provided me with friendly temporary homes during the international moves over the time this book was written: Gunnar and Krystyna Boehnert; my sister Jennifer Boehnert and Joel Davis, Lana Pitkin and Uroš Raičković; Mel MacDonald and Damian Bubicz; Adele Green; Jayne Quinton; Alex Burden; Rebecca Cheeham and Culam Nelson; Becky Heuberger and Piers Windsor; Ruth Winstanley and Phil Neale; Richie Kearon; Paul Chandler; Jenny Haynes; Brennan Linsley; Pj Damico; and Barley Massey. A special thanks to Alan King for the all the times he has been there for me over these years.

Finally, this book would not exist if I was not so inspired by my late mother, also Joanna Boehnert (born Bartlett), who died in a car crash in 2001. My love for her sits with more complicated feelings behind some of my motivations to write. I am compelled to articulate the tensions that troubled my relationship with her generation of feminists in ways that also acknowledge my gratitude to them and my respect for the work they did struggling against very different conditions than the ones I encountered. The world changed radically thanks to their work, but not nearly enough! Fortunately, many young people today have a more thorough analysis of power, exploitation and structural violence than those who had the privilege of inheriting a more stable climate and access to cheap fossil fuel energy. We will need better ideas than the ones that created the various problems we are facing. This book is dedicated to millennials and those who come after them.

Introduction: Within and Beyond Error

Whether or not we are interested in the environment or identify with the concept of being 'an environmentalist' each of us is entirely dependent on the air we breathe, the food we eat and the environment we inhabit for life. Despite this basic fact, statements about our connections with nature are often interpreted as platitudinous and widely dismissed. We have inherited a highly reductive intellectual tradition and anti-ecological worldview in profound denial of our fundamental interdependence with nature. We are embedded within non-human nature[1] and dependent on ecological systems for life, but our belief systems do not reflect this basic relationship. Consequently, the world we have designed is deeply unsustainable.

Fragmentary thinking is an obstacle to sustainability. Reductive attitudes towards knowledge cannot address problems associated with complex ecological systems – or social and economic problems for that matter. In response to this dilemma, ecological literacy provides an integrated foundation for the understanding of environmental problems and potential solutions. Unfortunately, ecological literacy has largely failed to spread across disciplinary boundaries in over two decades since the concept was first conceived, and it remains marginal in education, policy and practice. All too often it is absent in places where it is desperately needed – such as the disciplines responsible for the design of sustainable futures. In this book, I focus on what ecological literacy means for communication design, although many ideas will be relevant for other design disciplines and of interest to anyone concerned with designing sustainable ways of living.

Design serves a social and political function that is not always acknowledged within the design industry. While there are prominent movements in design working towards socially responsive practice, these efforts are hampered by the manner in which power relations are reproduced by design and the lack of analysis of these dynamics. Design is a practice that functions to mediate social relations. Typically, it reproduces the values and priorities of those who determine which design problems are to be addressed. The interests of powerful groups are manifested in design.

Communication design is strategically placed to be pivotal in the transformation of unsustainable ways of living. John Berger famously said, 'seeing comes before words' (2008 [1972], 7). Seeing is a way new ideas emerge. Since design can facilitate new ways of seeing (with its image making practices) design is a way new ideas are made tangible. In particular, communication design has unique properties that can nurture new perceptual and cognitive capacities supporting relational or ecological perception and ways of knowing. Communication designers are visualizers: they can construct future scenarios with imagery before it takes form. Designers are information mediators: they can organize information and help make it meaningful. Designers are also experts in subjectivity and regularly make us want to do things in new ways. Since communication is key to mobilizing responses at moments of contingency, communication designers have a significant role to play in responding to the current environmental crisis.

This is a book about design and communication (and especially communication design) in the context of converging ecological and social crises. On one level it is a book about how design influences and communicates; how it encourages us to feel how we feel; how it helps us know who we are and who we might become. It is also a book about how ideas, social relations and political structures influence how we treat our environment: what use we make of it and what care we take of it. Design negotiates the intimately intertwined space between self, society and the environment. What future we have (or don't have) will depend in many ways on what design concepts, strategies and methods we choose. In this book, I will advocate for ecoliterate design.

This book is organized in three parts:

Part 1: *Design* introduces the role of design in society. It examines the ways in which design functions to construct and reproduce social relations. It reflects on the potential of design to be a form of symbolic violence. It considers the tensions between design as a practice for social and ecological good and the priorities embedded in the design industry.

Part 2: *Ecology* explores the philosophical foundations of ecological thought, its history and exactly what it means to be ecologically literate. It describes how ecological theory informs the practice of design. It introduces ecological principles for design and the concepts of ecological perception and ecological identity. This is the largest part of the book.

Part 3: *Politics* examines how the interests of powerful groups are manifested in design and how design can obscure these interests. It deconstructs the narrative of 'doing good' in design by examining the effectiveness of various initiatives. It reviews how social marketing, the green economy, technofixes and information visualization all do political things.

Environmental problems are situated at the intersection of these three areas. The three spheres are inspired by Felix Guattari's *The Three Ecologies* (published in French in 1989 and translated to English in 2000) and Gregory Bateson's *Steps to an Ecology of Mind* (1972). Guattari, following Bateson, proposes that mental ecology, social ecology and environmental ecology are three realms that cannot be disconnected. In theory and practice we must work with the three ecologies (human subjectivity, social relations and the environment) simultaneously. For Guattari it is our failure to work with these realms (the mental, the social and the environmental) at once that creates contradictions and stunts efforts to address environmental problems. In *The Three Ecologies* he explains:

> So, wherever we turn, there is the same nagging paradox: on the one hand the continuous development of new techno-scientific means to potentially resolve the dominant ecological issues and reinstate socially useful activities on the surface of the planet, and, on the other, the inability of organized social forces and constituted subjective formations to take hold of these resources in order to make them work. (2000, 22)

In response to this dilemma, he calls for a theory of ecosophy, an 'ethico-political articulation' (Ibid., 19) that will consider the dynamics between the three ecologies. A new praxis to 'ward off, by every means possible, the entropic rise of a dominant subjectivity' (Ibid., 45). This work will be done by 'literally reconstructing the modalities of "group-being" ... through "communicational" interventions' for the modification and reinvention of the ways in which we live by 'the motor of subjectivity' (Ibid., 24). Keeping in mind that: 'There is at least a risk that there will be no more human history unless humanity undertakes a radical reconsideration of itself' (Ibid., 45), this work is necessarily hugely ambitious: 'In its final account, the ecosophical problematic is the production of human existence itself in new historical contexts' (Ibid., 24). The emergence of ecological sensibilities is a basis for transformative change.

Design is a practice well placed to respond to this call. Bateson first described how the ecological struggle is in the domain of ideas (1972, 495–505; Pindar & Sutton 2000, 11). Guattari called on all cultural practices 'in a position to intervene in individual and collective psychical proceedings' (2000, 27) to participate in this ethico-aesthetic project to nurture a new ecological subjectivity. This book moves the theory developed by Bateson and Guattari (also Plumwood, Orr, Sewall and others) closer to practice. As a theoretical foundation for ecologically informed design, the book explores the politics of why current design practice is stuck in the reproduction of unsustainability. In bridging social theory with ecological theory, the book addresses some historical suspicions between these two fields. The tensions between these two modes need to be encountered and conjoined. As we construct new ways of relating with each other and the ecological space we inhabit, within an order that works against just, socially equitable sustainability, we must live with contradictions embedded in the political system as we attempt to dismantle and transform structures and practices to enable viable, sustainable futures (Figures I.1 and I.2).

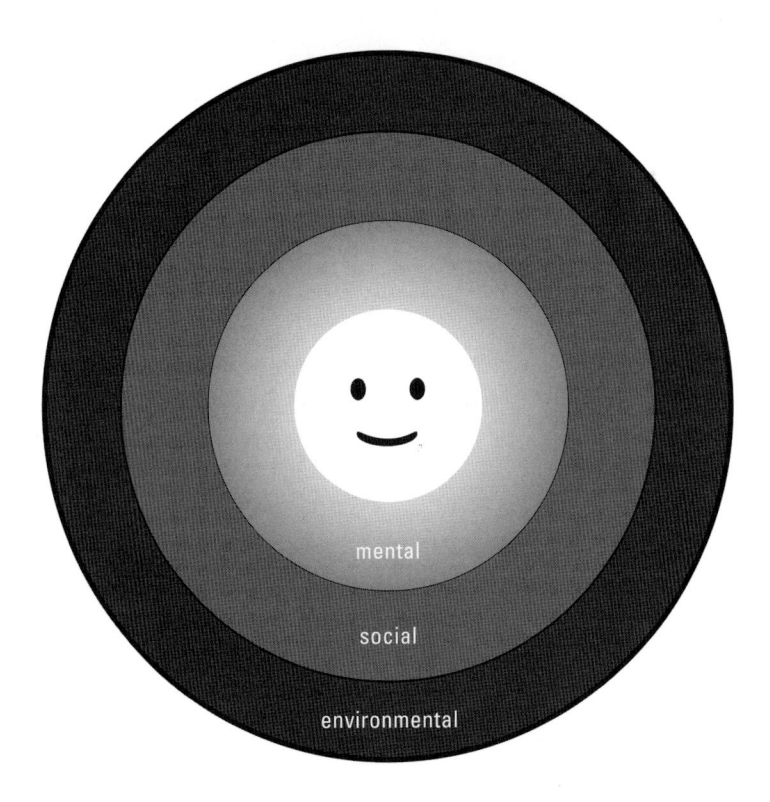

Fig. I.1 *The Three Ecologies A.* EcoLabs. 2017.

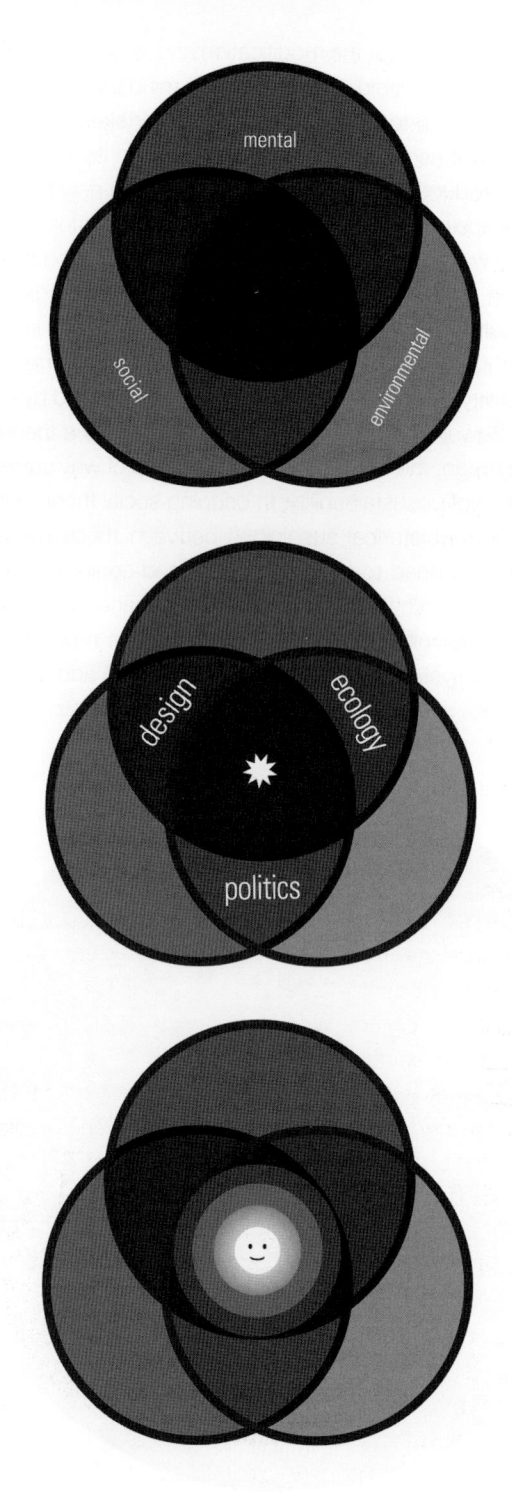

Fig. I.2 *The Three Ecologies (B, C & D).* EcoLabs. 2017.

The three ecologies are reinterpreted in this book (design, ecology and politics). The ordering in the title is not true to type in an ontological sense – but pragmatic: it reflects the order in which I introduce the theory of design, ecology and politics. Designers have expertise in influencing subjectivities. Design is a field that mediates the subjective realm (the mental ecology). Politics describes the ways in which the social realm is organized (the social ecology). Ecological theory considers human relations to the environment (the environmental ecology). In this book, I attempt to bring these three spheres together. This assembly of three ecologies is both a meeting of three domains (the environmental, the social and the mental) and a recognition that these are orders embedded in each other and need to be theorized simultaneously. Inclusive theory is a basis of good design.

This book bridges different traditions. Social theory reveals the social function of design including the ways ideas and ideologies are reproduced in the communication and the objects designers make. Ecological theory describes human relationships with the environment, including the ways in which design contributes to and reproduces unsustainable conditions. Political theory describes how particular constituencies control what is designed and how these power dynamics are concealed by designers (who are often unaware of the ideological work they perform). Located at the intersection of these fields, the book explores how designers participate in the construction of future realities by creating new ways of doing things. Revealing these dynamics creates new possibilities for transformative practice.

By linking social theory and ecological theory to design theory and practice, this work critiques the ways that the design industry – and even many sustainability discourses within the design industry, perpetuates current unsustainable development regimes. When design does engage with issues of sustainability, this engagement typically remains shallow due to a narrow basis of analysis in design theory and education. The situation is made more severe by design cultures that claim to have no politics. The supposedly 'neutral' designer is typically the most unaware of their own ideological assumptions and allegiances to power, normative values and the status quo. The political system in the UK (also North America and Europe) is a political system that is built with and for capitalist economic production with its neoliberal modes of governance. This system impacts how we live our lives and what is happening to the climate. Throughout this book I review what neoliberalism means for the environment and design.

Due to the scale of current challenges, new ecologically informed technologies and design practices (such as biomimicry, circular economy, renewable energy, life cycle analysis, etc.) must be incorporated into a larger project of political change. The insights from ecological theory can be applied to the political model itself. Design must be informed by both ecological literacy and social theory to help designers critically assess and navigate pathways to sustainability. (I provide the analytic tools for the more critical account of sustainability throughout this book.) Since sustainable futures depend on not only a technological innovation but also social and political change, designers must be aware of historical circumstances that have created unsustainable ways of living and the mechanisms of social change. Design education must broaden the scope and the depth of its analysis to attend to the complexity of contemporary problems. With these insights, design can be a transformative practice.

The design industry plays an important role in the creation of consumer desire – and of neoliberal sensibilities. While encouraging particular ways of thinking, design is also often involved in concealing the impacts of consumer capitalism and in obfuscating power relations. Situated at the hub of industrial production processes designers all too often have a cynical relationship to both consumerism and capitalism – as if there is no alternative. The neoliberal political project aims to abolish alternatives. Actually, there are plenty of options once we recognize the ideological work that is being done that destroys the visibility of other possibilities. Even noticing the ideological barriers to sustainability informs strategies of renewal. I will start with a hard look at the context in which we are acting.

An Overview of Earth System Sciences and Interpretations

Planetary Boundaries enable the relative stability of Earth Systems (ES) and create conditions amenable for civilization. The Planetary Boundary framework defines a 'safe operating space' for global societal development (see Figure I.3 and pp. 89–101). There are now four Earth Systems that have transgressed safe limits (climate change, biosphere integrity, biogeochemical flows and land-system change). Two of these (climate change and biosphere integrity) have the potential to drive the Earth into a new state – and not necessarily one that provides the conditions necessary for civilization. Biosphere integrity refers to pressures associated with biodiversity loss. The biosphere 'regulates material and energy flows in the ES and increases its resilience to abrupt and gradual change' (Steffen et

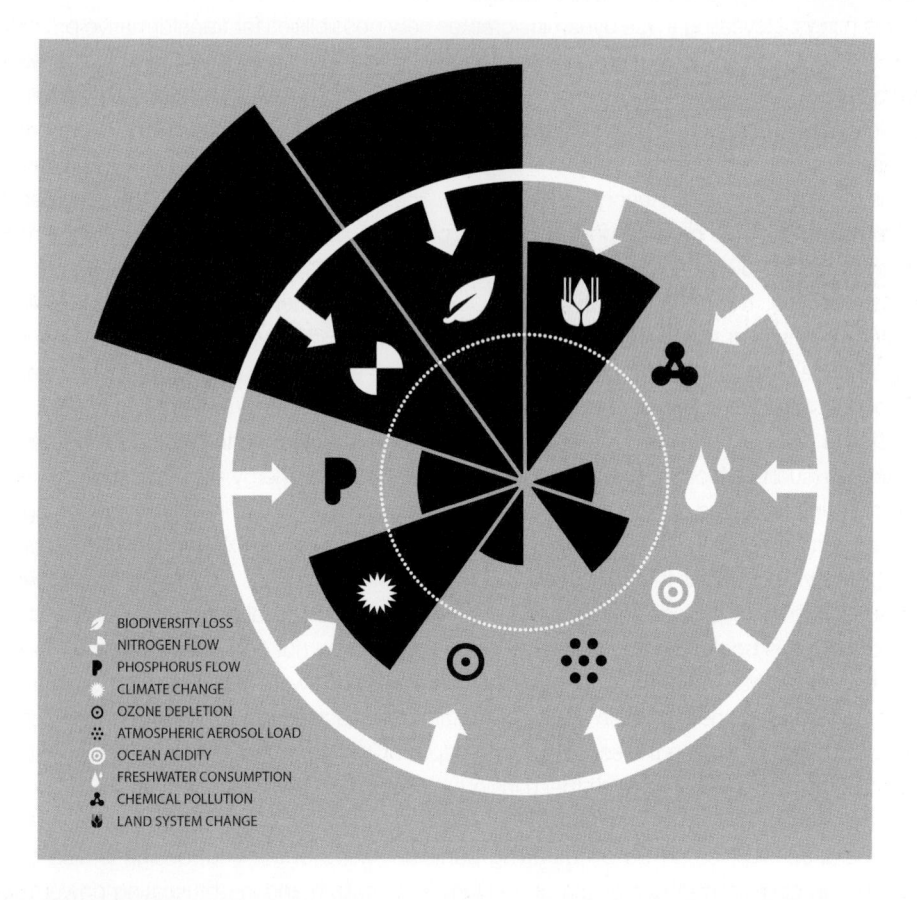

Fig. I.3 *Planetary Boundaries*, detail of *An Audit of Development*. EcoLabs + Tzortzis Rallis, Lazaros Kakoulidis, 2013.

al. 2015a, 736). Since all living things rely on their environment, biodiversity loss is impacted by each of the other Earth Systems. Current species extinction is now happening at a faster rate than any time since the last mass extinction event (65 million years ago). Humankind has triggered a Sixth Extinction (Kolbert 2014). Amphibians, the most endangered class of animal, are becoming extinct at 45,000 times the normal rate (Ibid., 17). While climate change tends to get more attention as the implications of destabilizing the Earth's climate system are extraordinarily perilous, all of the Earth Systems need to be considered concurrently to understand the scope of environmental harms and risks to humanity. 'Biosphere integrity' is an alienating term for something magnificent as the animals we share the Earth with, but the scientific term reminds us that other species are often essential actors in local ecosystems that enable the regenerative processes on which humans depend.

The risks associated with transgressing boundaries are severe: 'Anthropogenic pressures on the Earth System have reached a scale where abrupt global environmental change can no longer be excluded' (Rockström et al. 2009, 1). Climate change is already causing major disruptions in Earth Systems. Warming of the atmosphere and ocean system is unequivocal and associated impacts are occurring at rates unprecedented in the historical record. Climate change presents severe risks and these impacts will become increasingly expensive, difficult and even impossible to mitigate if action is not taken to dramatically reduce greenhouse gas emissions. If business-as-usual carbon dioxide emissions continue, between 21 per cent and 52 per cent of all species will be committed to extinction within the century (Hansen 2011). Tipping points in the Earth System can make change irreversible; 'recent greenhouse gas (GHG) emissions place the Earth perilously close to dramatic climate change that could run out of our control' (Hansen et al. 2007, 1). Despite these dangers, humanity's global carbon footprint has increased by eleven fold since 1961 (WWF INT 2010, 8) and continues to rise at rates higher than ever – global emission shot up by 5.9 per cent in 2010, the largest absolute increase ever (Klein 2014, 18 quoting Peters et al. 2011). International negotiations have not succeeded: in 2013 carbon dioxide levels were 62 per cent higher than in 1990 when the Intergovernmental Panel on Climate Change (IPCC) issued its first report and negotiations started (Klein 2014, 11). The 2015 UNFCCC COP21 'Paris Agreement' was presented as a success – but for the climate justice movement it was a dramatic public relations exercise in business-as-usual.

Industrial processes are dramatically impacting the Earth. Humankind has not yet learned how to use technology on scale in ecologically benign ways. Over the past forty years, the Living Planet Index (an indicator of the state of biodiversity) has fallen by 52 per cent (WWF 2014, 12). In 'less than two human generations, population sizes of vertebrate species have dropped by half' (WWF 2014, 4). At a global level, the yearly ecological footprint takes one and a half years of regenerative capacity or biocapacity to replace (WWF 2014, 9). Thus biocapacity continues to shrink while consumption rates continue to grow. Even the most basic analysis indicates the danger of this situation. Vastly disproportionate responsibility for global ecological problems lies with the rich: 'the world's richest 500 million people (roughly 7 per cent of the world's population) are currently responsible for 50 per cent of the world's carbon dioxide emissions' (Assadourian 2010, 6). The first causalities of degraded natural systems are the poor. Environmental problems are also social problems and need to be approached with environmental justice in mind. A whole system's audit of the current mode of development would reveal deep flaws with the current economic model (Figure I.4).

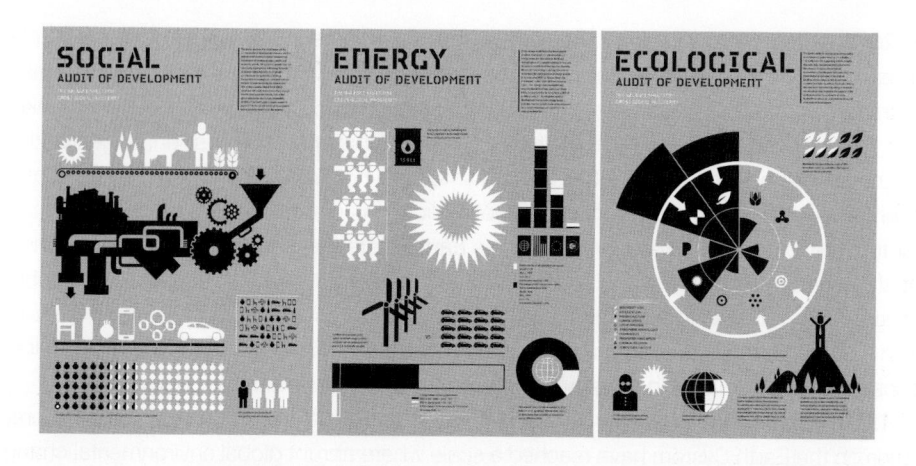

Fig. I.4 *An Audit of Development*. EcoLabs, Tzortzis Rallis and Lazaros Kakoulidis, 2013.

The vital signs of the planet are included here as they are the basic background knowledge necessary for responsible design. Regrettably, in many places there remains a complete disconnect between the environmental sciences and design – and therefore between these environmental circumstances and the practices advocated by design education and the design industry. It should be evident that design that is not sustainable is deeply unethical. The Millennium Ecosystem Assessment warns that: 'human activity is putting such strain on the natural functions of Earth that the ability of the planet's ecosystems to sustain future generations can no longer be taken for granted' (Assadourian 2010, 4). While many of us see non-human nature as having value outside of its worth for humankind, even if we have no regard for nature for its own sake, the de-stabilization of global ecological systems creates grave risks for humanity. Earth System change threatens to make all other long-term goals obsolete. The capacity of the ecological system to continue to provide favourable conditions for civilization is no longer assured. In the long term for everyone and immediate present for those in communities on the front line of climate impacts and other environmental harms: everything depends on the ecological context.

The Anthropocene

Due to the dramatic changes that humankind is inflicting on Earth Systems, scientists warn that we are now exiting the relatively stable Holocene epoch in which civilization developed. We are entering a new geological epoch, that of the Anthropocene, wherein humankind is responsible for altering the functioning of the Earth System (Crutzen & Stoermer 2000; Steffen, Crutzen & McNweill 2007; Zalasiewicz et al. 2015). This power over nature has not been accompanied with the foresight to use technological capacities wisely. In *Science*, Will Steffen et al. wrote:

> The relatively stable, 11,700-year-long Holocene epoch is the only state of the ES [Earth System] that we know for certain can support contemporary human societies. There is increasing evidence that human activities are affecting ES functioning to a degree that threatens the resilience of the ES – its ability to persist in a Holocene-like state in the face of increasing human pressures and shocks. (13 February 2015, 737)

There is debate in the scientific community on the starting date of the Anthropocene. Several dates have been proposed: the collision of the Old and New Worlds circa 1610 (Lewis & Maslin 2015); industrialization (1800–present) (Steffen, Crutzen & McNweill 2007); the 'Great Acceleration' following the Second World War (Zalasiewicz et al. 2015); the nuclear weapons detonation 1945–1961; and with persistent industrial chemical pollution (1950s–present). A potential date is that of the world's first nuclear bomb explosion, on 16 July 1945. Thereafter 'additional bombs were detonated at the average rate of one every 9.6 days until 1988 with attendant worldwide fallout easily identifiable in the chemostratigraphic record' (Zalasiewicz et al. 2015, 1). Nobel Prize winning atmospheric chemist Paul Crutzen, climate scientist Will Steffen and environmental historian John McNeil have proposed two distinct stages in the Anthropocene: the Industrial Era (from 1800 to 1945) and the Great Acceleration (from 1945 to the present) (Steffen et al. 2007). The Great Acceleration is evident in the charts in Figure I.5 where Earth System Trends all reveal dramatic changes mid-twentieth century.

Earth system trends

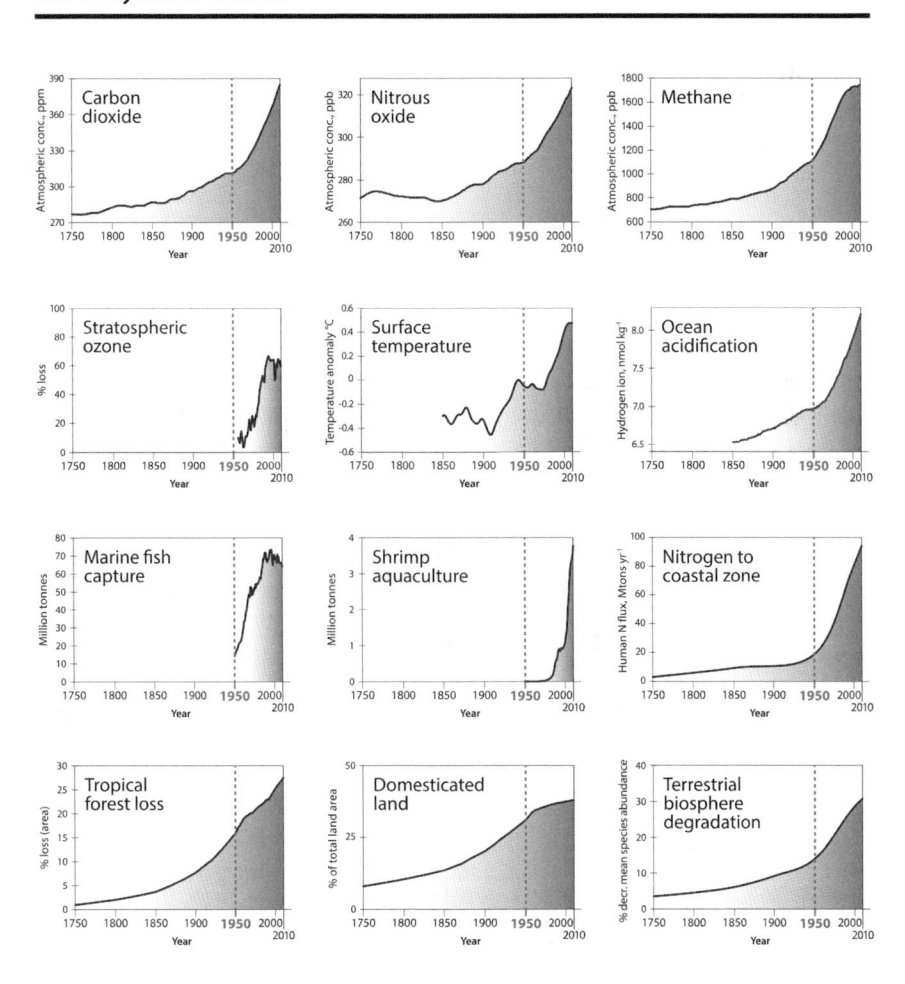

Fig. I.5 *Earth System Trends*. The Trajectory of the Anthropocene: The Great Acceleration. Steffen et al. 2015.

Simon Lewis and Mark Maslin's *Science* paper (2015) suggests that the death of over fifty million indigenous residents of the Americas in the first century after European contact constitutes the most significant marker of the new epoch. A proposal to formalize the 'Anthropocene' is under development. Whatever date is chosen, the Anthropocene's 'literal meaning – the "age of humans" – is either shocking or hugely flattering, depending on one's perspective' (Castree 2014, 235). But not all perspectives are equally well informed. Those who have the most knowledge of Earth system science are concerned and often alarmed.

The Capitalocene

The concept of the Anthropocene draws attention to severe environmental problems – but it also does other things. Jason W. Moore asks: 'Does the Anthropocene argument obscure more than it illuminates?' (2014, 4). The idea has been critiqued as uncritically importing Western rationality, imperialism and anthropocentrism and thereby narrowing humankind's options for developing sustainable alternatives (Moore 2014, 2015; Latour 2014; Haraway 2014, 2015). Specific activities are destabilizing climate systems and other planetary boundaries. The Capitalocene is a concept that asserts: 'the logic of capital drives disruption of Earth System. Not humans in general' (Solon 2014). Bruno Latour claims:

> The 'anthropos' of the Anthropocene is not exactly any body, it is made of highly localised networks of some individual bodies whose responsibility is staggering ... this dispersion of the 'anthropos' into specific historical and local networks, actually gives a lot of weight to the other candidate for naming the same period of geohistory, that of 'capitalocene', a swift way to ascribe this responsibility to whom and to where it belongs. (2014, 139)

Not everyone shares responsibility for ecologically destructive modes of development. Power and responsibility are concentrated on those who have the ability to influence industrial development and system structures. Dana Luciano notes that destruction of Earth Systems 'was not brought about by all members of the species it names' (2015, para. 16). The contradiction that is embedded in the name of the new epoch 'is precisely the problem it is now up to us to solve' (Ibid.). Distinguishing the specific systemic processes that drive ecological crises is key to identifying problems and constructing effective responses. The notion that all of us are responsible for the ecological crisis bolsters a particular narrative and serves the interests of those who would like to maintain business-as-usual. The name of the new epoch establishes a powerful metaphor and an associated framework that will help or hinder the struggle to mainstream sustainable design. As an alternative, the Capitalocene proposal highlights the role of capital accumulation in creating ecological crisis conditions:

> Our economic system and our planetary system are now at war. Or, more accurately, our economy is at war with many forms of life on earth, including human life. What the climate needs to avoid collapse is a contraction in humanity's use of resources; what our economic model demands to avoid collapse is unfettered expansion. Only one of these sets of rules can be changed, and its not the laws of nature. (Klein 2014, 21)

There is a certain model of development driving dramatic Earth System change. Humans in general have other options beyond the ones we have currently constructed. For this reason, it is important to be specific about what type of development we *Anthropos* of the Anthropocene pursue.

The Ecocene

Beyond the hubris of the Anthropocene and the critique of the Capitalocene, new ways of understanding and organizing social and ecological relations are emergent. Critiquing the ideas, politics and technologies that have contributed to ecological crisis is only a starting point for the work to be done creating ecologically viable ways of living. Design theorist Rachel Armstrong coined the concept of an 'Ecocene'. The Ecocene shifts focus from the problems to the solutions: 'There is no advantage to us to bring the Anthropocene into the future … The mythos of the Anthropocene does not help us … we must re-imagine our world and enable the Ecocene' (Armstrong 2015). The challenges today are hyper-complex but we must bring 'the great dithering' to an end and catalyze a transition (Ibid.). For Armstrong, now is an exciting time to be a designer. The Ecocene has yet to be designed. Its emergence depends on a new understanding of human–nature relations and new types of development and design that emerge from this perspective. New ecologically informed ways of thinking and living must be generated. The transformative Ecocene describes a curative catalyst for cultural change necessary to survive the Anthropocene. The emergence of the Ecocene depends entirely on what we do now.

This brief overview cannot do justice to either the scope of environmental problems or the difficulty in protecting ecosystems while also committing to social and environment justice within the current economic paradigm. The environment is the basis for prosperity and a foundation for social justice. Addressing its problems is a basic imperative. We require strong moral sanctions against life destroying industrial development. Until these can be created, as long as we live in a society with capitalism and a centralized state, we need robust mechanisms in the public sector to monitor, regulate and transform industry along with strong laws against ecocide. These should not be seen as end goals but rather means of coping with the current problems while we design ways of living within and beyond the error embedded in this political system so replete with contradictions. Design is a practice that can help make this happen.

Note

1. The term 'non-human nature' accentuates the fact that humans are also part of nature.

Part 1 Design

, # Part 1 Design

1 Design Theory 101

Typically design is described as a problem-solving practice that meets human needs and desires with the creation of new artefacts, communication, fashion, buildings, spaces, services and systems. To do this work effectively, design must also be a problem diagnosis practice that draws on a wide range of disciplinary traditions to interpret and analyse issues under investigation and come up with creative ways forward. Design is also a practice that mediates social relations by creating the artefacts that enable and enrich everyday lives. It can be understood as a decentralized social and technical knowledge building practice where designers self-organize to produce innovative solutions and new types of order. The ways by which designers solve problems are emergent processes. Socially responsive design approaches must first accurately identify the problems to be solved, and then intensify the positive potential of design while circumventing its more manipulative tendencies.

In the tradition of ecological literacy advocate David Orr, I am theorizing design as 'a large and unifying concept – quite literally the remaking of the human presence on Earth ... [as] how we provision ourselves with food, energy, material, shelter, livelihood, transport, water and waste cycling' (Orr 2007, unpaginated). At its best, design is an integrative transdisciplinary field of conceptual and applied tacit knowledge that bridges theory and action in pursuit of practical outcomes. Designers have always addressed social problems as well as economic ones and over recent decades pioneers have opened the scope of design, often involving a shift from designing artefacts and products to co-designing and facilitating new processes, services, systems and ways of living. These movements have become more pronounced as more people recognize the scope of environmental problems and the need to create more effective tools and strategies for engaging with complexity. In response to climate change and other socio-ecological problems, radically new ways of living must be created. Design is primed to function as a facilitator of social and technological change once it transcends and helps transform the systemic priorities of the current development framework.

Design is often concerned with influencing audiences and users in order to change both ideas and behaviours. Designers create change-making artefacts and communications that encourage people do new things. Design is effective at change-making because instead of telling people what to do and think, it creates new communication, metaphors, tools and techniques to enable people to see the value in new ways of doing things. Design shows rather than tells. Sustainable design pioneer Buckminster Fuller said: 'If you want to change how someone thinks ... Give them a tool, the use of which will lead them to think differently' (Fuller quoted in Ehrenfeld 2008, xiv). Graphic design guru Edward Tufte has similar advice: 'To prove a point a designer should not argue, but demonstrate with data, with tools with techniques' (2010). Design is a means of nurturing new ways of thinking as a prerequisite to new behaviours. The various design disciplines share an orientation for enabling new ideas, attitudes and behaviours. With this in mind, design can also function as a means for revealing ecological relations and creating sustainable alternatives.

Design can be a subtle art that influences emotions, feelings and sensibilities. It is a practice that offers what technocratic approaches to sustainability lack: a means of making change attractive and emotionally satisfying. Its effects can be immediate, perceptual and non-rational. Good design makes artefacts and communication that 'feel right'. Aesthetics offer pleasure and meaning. Auschwitz concentration camp survivor Viklor Frankle is not alone in suggesting that our greatest motivation is 'not to gain pleasure or avoid pain but rather to see a meaning' (1959, 115). Communication designers are experts in working with identity, spirit and purpose. A good designer gives a space, object, process or experience delight, beauty and/or dignity. The ability of communication design in particular to give pleasure, construct meaning and illicit desire is why it is so highly valued in advertising.

Design uses tacit knowledge and creative processes to address the needs and desires of consumers as interpreted by those who determine what is to be created. Since design is necessary for innovation and this inventive process is integral to capitalism, design and capitalism have a cosy relationship. Design is the practice and profession that creates the seductive material goods and advertising that makes capitalism so exciting – and so destructive. In this mode, design drives conspicuous consumption, disciplines consumers, glorifies corporate power and also obscures these processes.

Design as the Mediation of Social Relations

Designers and other cultural producers mediate social relations with the cultural work they perform. By spreading new ideas and with the creation of new objects and new ways of doing things, design helps construct relationships between people and between people and the environment. Design mediates social relations with new communication, objects, services and environments that shape how we live and the ways we experience and relate to each other and to the material space we live in. Design can also work to normalize new circumstances and relationships by making ideas, artefacts and spaces seem acceptable even when grave injustices and ecological harms are done.

Since we need communication, artefacts, architecture and services to survive, most of us are quite willing to allow designed communication, objects, spaces and systems to mediate our relationships. Consumers have the right to choose between products that appear to give us a variety of options so it can seem as if we are making decisions on how we will live. Since ecological literacy is still nascent, the narrow limits of the design options available are not always apparent to everyone. To those paying attention to the gravity of environmental problems and the not unrelated social problems, the design options on offer are woefully inadequate. Since we are all controlled to some degree by the world that is designed for us and impacted by industries whether we buy their products or not – the options matter.

To many of us concerned with the environment, it appears as if we are locked into systems that were chosen for us. Our trip to work is destroying the climate (unless we use a bicycle or public transport where available). The computers we use are made in factories with poor working conditions with processes that produce vast amounts of toxic waste. In the United States especially, consumers do not always have the right to know how the circumstances under which their food is produced (how animals are treated, how crops are grown, the safety of various toxins, etc.). In the design industry, those who decide what is designed and how it will be designed make decisions that will affect not only all of us (although impacts are unequally distributed) – but also those who come after us and the rest of the non-human natural world (who have no voice to protest the life-destroying conditions that are created).

While design plays a decisive role in mediating social relations, the important decisions about what will be designed are made far up the chain of command (by those who determine production policies and construct the economic framework). The political economy of design determines what is designed and whose interests are served. From this analysis, it might seem as if individual designers have little power. Yet because design is a field of practice that plays a powerful role in reproducing the dynamics of the system, there is latent power that is not being utilized due to 'remote controlling' (Guattari 2000, 26) through ideology embedded in design (among other places). The system would fall apart instantly if enough of us stopped ascribing to norms that enable it. It is the subjective grip of the values of the dominant order that designers reproduce.

Political theorist William Connolly describes a situation wherein 'a few corporate overlords monopolize creativity to sustain a bankrupt way of life … in which the ideology of freedom is a winnowed to a set of consumer choices between preset options' (2013, 79–80). Design within the current economic system is a process that reinforces those sensibilities encouraged by the neoliberal ruling order. Radical individualism, enterprising subjects, competitive relations and an idealized version of market autonomy are parts of an ideology internalized and embedded in Euro-American life (Connelly 2013; Dardot & Laval 2013; Gilbert 2014). Since this worldview is based on incomplete premises (as I describe in Part 2), the results are disastrous not only for the Earth but also for the vast majority of people on the planet. As a consequence of escalating inequality and desperate conditions, military prison and security budgets skyrocket as social crises proliferate.

The eco-social crisis is a crisis of values, priorities and associated subjectivities that are reproduced by culture, education and system structures. Communication designers can reproduce cultural norms and represent the non-human natural world as a space open to continued exploitation – or they can reveal ecological circumstances and nurture ecological ways of knowing. Ultimately, the options open to designers are determined by not only their employers' demands but also by any work they do beyond the capitalist order. In theory, designers could transform the world by making ecologically literate and socially responsible communication, products and spaces. In practice this will not happen within the workplaces of the vast majority as long as the model of development de-prioritizes ecological and social values – and design practice follows suit.

The Political Economy of Design: Design as Emergent Order

A social order refers to the structures and practices that maintain a society. All social orders (including those of ants and bees) develop patterns and processes for allocating and distributing resources (Varoufakis 2012, para. 2). The idea of emergent order, also known as spontaneous order, is that in the absence of a centralized command structure, conventions emerge to organize social activities. Order emerges in a manner based on the people's intentions as opposed to authoritarian hierarchies. Friedrich von Hayek used this theory to promote his brand of (supposedly) decentralized capitalism[1] that has now evolved into neoliberalism. The results of this experiment have been harsh (for the vast majority of people and for the environment) for two major reasons. The first is that the project narrowed the focus of the system exclusively to the priorities of the profit-oriented market while ignoring the interests of the social and ecological orders. The second is that the project has no safeguards against corporate power merging with the state creating even greater types of authoritarian regimes (Varoufakis 2012, para. 13). What is interesting about neoliberalism is its recognition of the potential of self-organizing systems and its (supposed) attempt to manage complexity by decentralizing power.

What is wrong and potentially catastrophic about this project is that it elevates the profit-seeking market to be the primary way of mediating social relations and dismisses the priorities of the orders that support and are the context of the economic order. With its exclusive focus on profit-making markets and using pricing mechanisms to determine the extraction and allocation of resources, this system undermines both the social and the ecological orders on which the market depends. It exploits and even destroys its own contexts. This is the essential contradiction in capitalism. With every failure caused by this contradiction, more authoritarian mechanisms are used to attempt to maintain the current social order in an increasingly tenuous situation.

The concept of emergent order itself is a useful theoretical model for understanding how design works in society. It considers the self-organizing powers of systems and the creative capacities within local decentralized networks. The phenomenon of emergence is significant for design because it describes a process of decentralized self-organization that results in the creation of entirely new properties. In design, some examples of emergent properties are designers' own new relational and visuo-spatial capacities that enable greater contextual understanding and new abilities to respond to complex levels of causality. These abilities support design's capacity to attend to sustainability challenges.

Design evolved from the tradition of craftsmanship wherein persons held practical skills for making new artefacts. It continues to be a discovery process that occurs in decentralized spaces as individual designers use tacit skills, strategies and tools to address problems. Accumulated tacit knowledge is used for the purposes of solving increasingly complex problems. For example, a communication designer has tacit knowledge manifested as drawing skills, developed through practice and study of master draftspersons. These skills can help a community understand proposals for an architectural development through a series of visualizations. Design can be understood as a process of embodying social rules in new communication, artefacts and spaces, thereby reproducing social rules and social relations while potentially solving problems – while sometimes making entirely new problems.

Design is uniquely positioned to engage with problem-solving in a dynamic process of moving from theory to practice and moving between disciplines and sectors to facilitate transdisciplinary actions. It is a field of practice engaging with increasing levels of complexity. As communication media change, humankind develops new communicative capacities (McLuhan 1964; Rushkoff 1996). Within an increasingly visual culture, the emergence of greater systemic thought is evident (Barry 1997; Horn 1998). Visual and digital communication is increasing human capacity for greater understanding of complexity and dynamics systems (Rushkoff 1996; Chabris & Kosslyn 2005) (see Chapters 9 and 10). Human capacities for negotiating complexity are enhanced by the visuo-spatial intellectual abilities of designers, the technologies they create and new design practices that emerge from this congruence. Emergent relational capacities support a new understanding of networks and complex levels of causality. As these abilities evolve our potential collective capacities to attend to sustainability challenges are enhanced.

Emergent cognitive capacities (such as critical, reflective, relational and ecological thought, and systems thinking) theoretically have radical implications for the design of innovative, prosperous and sustainable ways of living. The dissemination of knowledge within the design industry increases problem-solving capacities. Design evolves through knowledge sharing of successful design interventions. Good design solutions are imitated, successful strategies copied and these new projects can create more effective solutions. Despite the potential in the processes described above, efforts to use design to solve humanity's most urgent problems are not currently de-escalating ecological or social crises on scale.[2] The theory of emergent order helps to explain why. Instead of harnessing new abilities to address social and ecological problems, the corporate and industrial use of design typically harnesses the vision, skills and capacities of designers to serve its own goals, that is, the creation of economic profit. As the technological and industrial capacities of civilizations become more powerful, designers are increasingly implicated with ecologically and socially harmful (put profitable) design activities. Design straddles the borders of various systems: the economy (a physical and socially constructed system), society (a biological and socially constructed system) and the ecological system (a biological and geophysical system that is now impacted by the societal and economic orders). The current economic system, capitalism, has been constructed to respond to only signal: profit. It has not been designed to take other systems into account. Macroeconomics is currently based on the denial of biophysical reality. I describe problems associated with this dynamic in the political economy of design at length in Chapters 3 and 13.

The Role of Design in Capitalism

The critique of design as a facilitator of conspicuous consumption and glorifier of corporate power is not new to design theory. The pioneer sociologist Thorstein Veblen (1857–1929) identified conspicuous consumption as a dynamic wherein public displays of consumer goods signify social class and prestige. It's now been over fifty years since Ken Garland published the 1st edition of the 'First Things First' manifesto in 1964 (signed by twenty other visual communicators). IN THIS TEXT Garland asked (quite politely) not for 'the abolition of high pressure consumer advertising: this is not feasible' – but: 'a reversal of priorities'. The manifesto was revised in 1999 by Garland and thirty two other designers who published a new version in *Eye Magazine* and *AIGA Journal of Graphic Design*:

> Designers who devote their efforts primarily to advertising, marketing, and brand development are supporting, and implicitly endorsing, a mental environment so saturated with commercial messages that it is changing the very way citizen-consumers speak, think, feel, respond, and interact. To some extent we are all helping draft a reductive and immeasurably harmful code of public discourse. (Garland et al. 1999)

This new text also proposed 'a reversal of priorities in favour of more useful, lasting, and democratic forms of communication' (1999). By the turn of the millennium the critique was well established – although the problems with over consumption and the resulting pollution continued to grow. Since this critique is such a popular one in design theory, it is worth asking why more progress has not been made.

Victor Papanek contributed a more demanding text in his seminal 1971 book *Design for the Real World: Human Ecology and Social Change*. In this text, Papanek states:

> There are professions more harmful than industrial design, but only a very few of them. And possibly only one profession is phonier. Advertising design, in persuading people to buy things they don't need, with money they don't have, in order to impress people who don't care, is probably the phoniest field in existence today. (revised version 2nd edition 2000, ix)

Four years later, in the essay 'Edugraphology: The Myth of Design and the Design of Myths' (1975) he developed a critique of communication design and education by describing how both were dedicated to the following six discernable directions:

1. To persuade people to buy things they don't need with money they don't have to impress others who don't care.
2. To persuasively inform about the class-merits of an artifact, service, or experience.
3. To package in a wasteful and ecologically indefensible ways, artifacts, services or experiences ...
4. To provide visual delight or visual catharsis to those classes taught to respond 'properly.'
5. To undo with one hand what the other has done. (Anti-pollution posters, anti-cigarette commercials).
6. To systemically research the history, present, and future practices in the fives fields listed above.

(Published in Papanek 1999, 252–253)

In response to the 'myth' that designers solve problems, Papanek explains how problem-solving works in the design industry by focusing on certain types of solutions: 'a graphic designer "solves the problem" of advertising rail-travel as ecologically saner than automobile-travel, but at the cost of neglecting walking or biking, *and in doing so diminishes the choices people can make*' (1999, 254). Papanek lays a foundation for socially responsive design by asserting that design will only be ethically beneficial for the majority once designers focus on real needs, rather than manufacturing needs for those with money to spend (and normalizing the power dynamics around increasing inequality). The text sharply critiques the quest for profit but stopped short of a deeper analysis of the structures and systems that determine that profit must always be prioritized.

The twentieth century saw a transformation in how people live that was very much constructed by design. Pollution (including climate change) is a result of boardroom decisions by industrialists and made into a reality by the practical work of many professions including design. For example, planned obsolescence is a strategy of deliberate and entirely unnecessary ecological harm. It refers to products that are intentionally designed and manufactured with artificially limited useful life, wasting resources and creating more pollution than necessary. Vince Packard popularized the critique of planned obsolescence in the 1960 book *The WasteMakers*. Over fifty-five years later, these unscrupulous design strategies are still happening – and proliferating.

Consumer goods are advertised using all the insights of psychology and the social sciences. Edward Bernay, the nephew of Sigmund Freud, applied psychological insights to public relations and advertising starting in 1920s. He created an array of highly 'successful' campaigns that achieved goals such as making smoking popular with young women amongst other non-benign accomplishments. Design theorist Jonathan Chapman stresses: 'consumers of the 1900s were not born wasteful, they were trained to be so by the sales-hungry teaching of a handful of industries bent on market domination' (2005, 9). Designers are needed to facilitate these processes by designing new products and making them appealing to new audiences with advertising.

Despite the fact that the decisions about what consumer goods will be available and promoted are made industry leaders, it serves the interest of industry to shift the blame. The 'responsibilization' discourse is one of devolved responsibility wherein the ecological crisis is blamed on ordinary people, as if it was their greed that made the current ecological crisis inevitable. This discourse is only possible due to the ways that corporate power has been camouflaged. In 'Notes for the New Millennium. Is the Role of Design to Glorify Corporate Power?' (first published in 1990) Stuart Ewen accuses design to be a practice that

> regularly masks destructive patterns within our society. In day-to-day practice, with little or no self-examination, commercial design routinely aestheticizes, or renders beautiful, hazardous ideas about the use of environmental resources, the nature and concentration of power, and about the ordering of value by which we live. In the process, tendencies which threaten the interests of human survival are transformed into icons of 'the good life'. (2003, 193)

According to Ewen's text, the environmental crisis is a result of a system wherein 'engineered waste and obsolescence are prime stimulants of "economic health" ' (Ibid., 193). Design facilitates this situation wherein 'the range of human possibility is being incrementally narrowed to that which is for sale' (Ibid., 194). Design puts a 'gloss on social inequity and limit[s] utopia in the marketplace' and 'serves to aestheticise and validate waste, anti-democratic forms of power and the primacy of surface or substance' (Ibid., 194). Where design functions in these dark ways, it does so because the 'imperatives' of the economic context which prioritize short-term profits over long-term ecological sustainability.

Meanwhile much shortsighted environmental analysis focuses on a critique of consumption rather than a critique of production (and in particular the production of consumption). This is where designers stand in a pivotal position in the transition to an ecological future. Since design is necessary for the production of either ecologically harmful or beneficial ways of living – ecologically literate designers could make a difference. This work must go far beyond design strategies for nudging or budging behaviour.

While the analysis above clarifies problems, it does not necessarily support the capacities and the agencies necessary to address these problems. Designers of all people should realize that social change is never the result of issuing manifestos – or writing essays and books for that matter. (In my own defence, I will add that good theory is necessary for good practice). Neither system structures nor behaviours change by simply critiquing a situation and announcing new values. Theory itself can be transformative on a personal level but awareness alone will not solve our problems. Designers and design theorists are responding to the socio-ecological problems described above with socially responsive design strategies.

Socially Responsive Design Approaches and Strategies

As the ways in which design can be used to address social problems has become evident the discipline has evolved quickly over the past two decades to develop more effective socially responsive strategies. Sustainable design theorist Terry Irwin describes this 'tremendous change' as a prompted by two developments: (1) other fields and disciplines are adopting design tools and methods to find, frame and visualize problems; and (2) a heightened awareness of wicked problems and the need for interdisciplinary problem-solving methods to address these problems (2015, 230). Irwin describes three areas of design research, practice and education: Design for Service, Design for Social Innovation and Transition Design (Ibid., 230) (see Figures 1.1 and 1.2). These three approaches 'can be situated along a continuum in which spatio-temporal contexts expand and deepen' (Ibid., 229). These design approaches vary according to the theories of change, assumptions, 'horizons of time, depth of engagements, and alternative socioeconomic and political contexts' (Ibid., 231) (which become more expansive along the continuum). In this section, I will add a fourth approach to this model. Design Activism is the use of design in ways that are more explicitly politicized than the three approaches described in Irwin's continuum.

Design for Service takes a systems approach to problem-solving and expands the scope of design from products and communication to the design of services, experiences and interactions. According to Irwin, Design for Service is now a mature discipline that works within existing socio-economic and political paradigms (Ibid., 231). Service design is well established in design education and industry.

The Design Council created the RED group in 2004 to bring design thinking to public services in the UK. The team developed the concept of transformation design as an interdisciplinary methodology for socially engaged design practice. The project came to a premature end two years after it was created, but in the short time that they existed, the team developed a framework for Transformation Design that was influential.

> **Features of Transformation Design** – RED Group, The UK Design Council
> 1 Defining and redefining the brief
> 2 Collaborating between disciplines
> 3 Employing participatory design techniques
> 4 Building capacity, not dependency
> 5 Designing beyond traditional solutions
> 6 Creating fundamental change
>
> (Burns et al. 2006, 20–21)

A more enduring approach can be found in the work of New York based Humantific and NextDesign Leadership Network which uses design methods to support processes it calls 'sensemaking' and 'changemaking'. They describe Design 1.0–4.0 as four phases in designing for different levels of complexity:

> **Humantific's Design 1.0 – 2.0 – 3.0 – 4.0**
> Design 1.0 – Artefacts and communications (traditional design)
> Design 2.0 – Products/services design
> Design 3.0 – Organizational transformation design (bounded by business or strategy)
> Design 4.0 – Social transformation design (complex, unbounded) (Humantific 2010)

A Continuum of Design Approaches

Mature discipline
Design for Service

Design within existing socio-economic & political paradigms

Solutions reach users through many 'touch points' over time through the **design of experiences**. Solutions are based upon the observation and interpretation of users' behavior and needs within particular contexts. Service design solutions aim to provide profit and benefits for the service provider and useful and desirable services for the user (consumer). Solutions are usually **based within the business arena and existing, dominant economic paradigm.**

Developing discipline
Design for Social Innovation

Design that challenges existing socio-economic & political paradigms

Design that **meets a social need more effectively than existing solutions**. Solutions often leverage or 'amplify' exsiting, under-utilized resources. Social innovation is a 'co-design' process in which **designers work as facilitators and catalysts** within transdisciplinary teams. Solutions benefit multiple stakeholders and empower communities to act in the public, private, commercial and non-profit sectors. **Design for social innovation represents design for emerging paradigms and alternative economic models, and leads to significant positive social change.**

Emergent discipline
Transition Design

Design within radically new socio-economic & political paradigms

Refers to design-led societal transition toward more sustainable futures and the reconception of entire lifestyles. It is based upon an understanding of the **interconnectedness and inter-dependency of social, economic, political and natural systems**. Transition Design focuses on the need for **'cosmopolitan localism'**, a place-based lifestyle in which solutions to global problems are designed to be appropriate for local social and environmental conditions. Transition Design challenges existing paradigms, envisions new ones, and leads to radical, positive social and environmental change.

→ Scale of time, depth of engagement, and context expand to include social & environmental concerns

© School of Design, Carnegie Mellon University, 2014

Design for Interactions

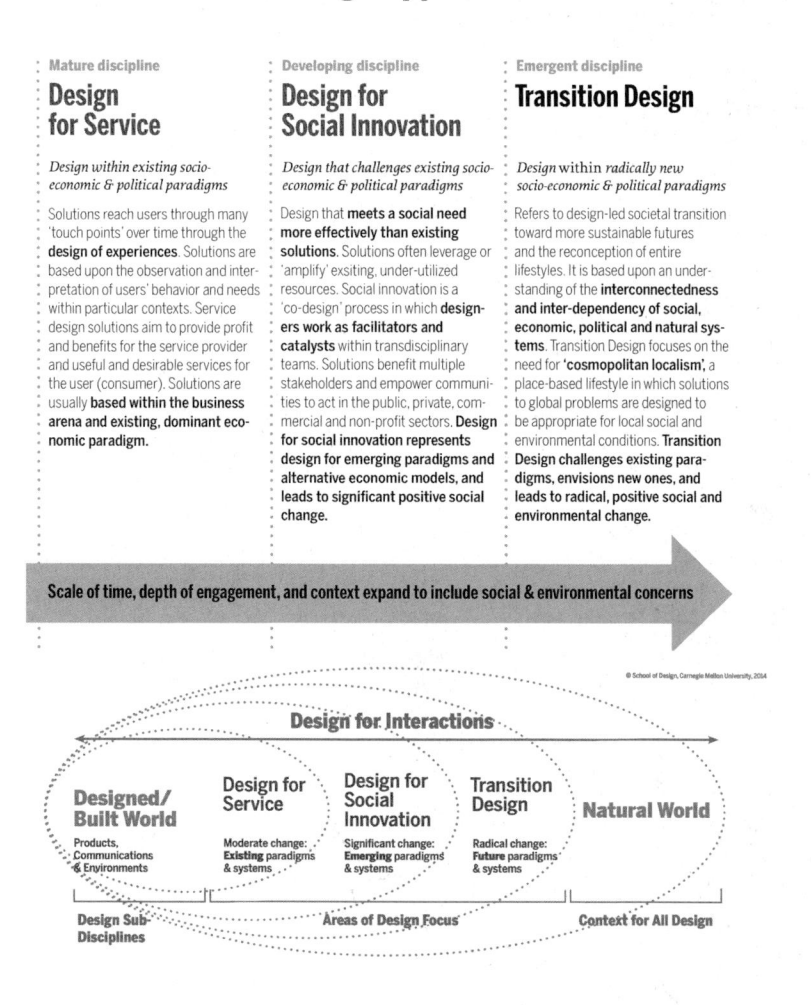

Designed/Built World	Design for Service	Design for Social Innovation	Transition Design	Natural World
Products, Communications & Environments	Moderate change: Existing paradigms & systems	Significant change: Emerging paradigms & systems	Radical change: Future paradigms & systems	

Design Sub-Disciplines | Areas of Design Focus | Context for All Design

Fig. 1.1 *A Continuum of Design Approaches* (Irwin et al. 2015).

The 1.0 – 2.0 – 3.0 – 4.0 framework mirrors Irwin's conceptualization of the expanding scope of design practices. Humantific's 4.0 expands design practice into social innovation beyond with visual sensemaking as a means of engaging with complex problems. A 'complexity navigation toolkit' is used with the assumption that change is dependent on new ways of seeing, understanding and thinking. Learning is a prelude to action. Visual strategies are used to re-access and reframe problems to be solved by design.

Design for Social Innovation further expands the contexts and objectives of design. It is 'usually situated within social and community contexts engagements are ideally longer, and solutions begin to challenge existing socioeconomic and political paradigms' (Irwin 2015, 230–231). Designers draw on a wide range of disciplinary traditions to inform an expanded practice.

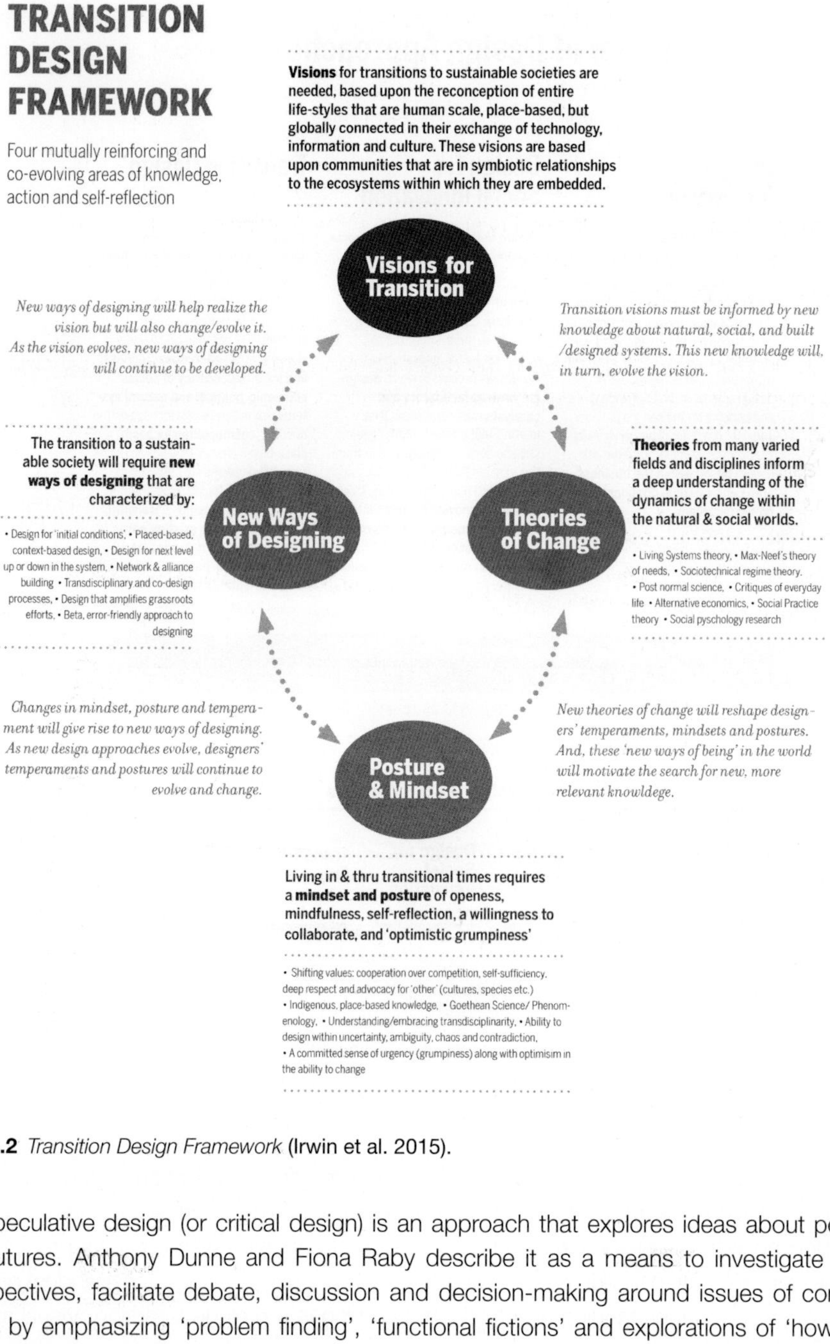

TRANSITION DESIGN FRAMEWORK

Four mutually reinforcing and co-evolving areas of knowledge, action and self-reflection

Visions for transitions to sustainable societies are needed, based upon the reconception of entire life-styles that are human scale, place-based, but globally connected in their exchange of technology, information and culture. These visions are based upon communities that are in symbiotic relationships to the ecosystems within which they are embedded.

New ways of designing will help realize the vision but will also change/evolve it. As the vision evolves, new ways of designing will continue to be developed.

Visions for Transition

Transition visions must be informed by new knowledge about natural, social, and built /designed systems. This new knowledge will, in turn, evolve the vision.

The transition to a sustainable society will require **new ways of designing** that are characterized by:

• Design for 'initial conditions', • Placed-based, context-based design, • Design for next level up or down in the system, • Network & alliance building • Transdisciplinary and co-design processes, • Design that amplifies grassroots efforts, • Beta, error-friendly approach to designing

New Ways of Designing

Theories of Change

Theories from many varied fields and disciplines inform a deep understanding of the dynamics of change within the natural & social worlds.

• Living Systems theory, • Max-Neef's theory of needs, • Sociotechnical regime theory. • Post normal science, • Critiques of everyday life • Alternative economics, • Social Practice theory • Social pyschology research

Changes in mindset, posture and temperament will give rise to new ways of designing. As new design approaches evolve, designers' temperaments and postures will continue to evolve and change.

Posture & Mindset

New theories of change will reshape designers' temperaments, mindsets and postures. And, these 'new ways of being' in the world will motivate the search for new, more relevant knowldege.

Living in & thru transitional times requires a **mindset and posture** of openess, mindfulness, self-reflection, a willingness to collaborate, and 'optimistic grumpiness'

• Shifting values: cooperation over competition, self-sufficiency, deep respect and advocacy for 'other' (cultures, species etc.)
• Indigenous, place-based knowledge, • Goethean Science/ Phenomenology, • Understanding/embracing transdisciplinarity, • Ability to design within uncertainty, ambiguity, chaos and contradiction,
• A committed sense of urgency (grumpiness) along with optimism in the ability to change

Fig. 1.2 *Transition Design Framework* (Irwin et al. 2015).

Speculative design (or critical design) is an approach that explores ideas about possible futures. Anthony Dunne and Fiona Raby describe it as a means to investigate new perspectives, facilitate debate, discussion and decision-making around issues of controversy by emphasizing 'problem finding', 'functional fictions' and explorations of 'how the world could be' (2013, 2). Dunne and Raby explicitly link this approach to the dilemma famously described by Fredric Jameson: 'it is now easier for us to imagine the end of the world than an alternative to capitalism'. They argue that 'alternatives are exactly what we need' (Ibid., 2) with 'more pluralism in design, not just of style but of ideology and values' (Ibid., 9) since 'if our belief systems and ideas don't change, then reality won't change either' (Ibid., 189). Theoretical and practical work in this tradition has opened up space in design for experiment, prefigurative practices and the envisioning of alternatives – while also creating links to critical thought. In practice, the depth of critically engagement often depends on designers' efforts to encounter controversy, and these engagements are not always evident.

Tony Fry and Anne-Marie Willis theorize redirected practice as a directional change in design from a practice that is 'elemental to the unsustainable' to a 'pathfinding means … to creating far more viable futures' (Fry 2009, 12, 7). Redirected practice aims to end design's complicity with the creation of defuturing conditions. Fry states that this redirection can only come about when 'designers place the current needs of the market in second place to the political-ethical project of gaining sustain-ability' (2009, 46). Like many activists who struggle to embed social and ecological values in design, Fry and Willis theorize the role of the market in perpetuating unsustainability and the role design plays in the dynamics of the market economy. They claim that design must facilitate the creation of new social institutions. As part of this re-direction in design, designers must learn to approach complexity without negating the complex (Willis 2010, 2). Speculative Design and Redirective Practice both open up the possibilities for design outside of its current roles and set the stage for deeper engagements with change-making practices.

Transition Design has recently emerged as an approach that 'takes as its central premise the need for social transitions to more sustainable futures and argues that design has a key role to play in these transitions' (Irwin et al. 2015, 1). Gideon Kossoff first wrote about Transition Design in his PhD thesis (2011, 5–24) where he described 'a design process that requires a vision, the integration of knowledge, and the need to think and act at different levels of scale, and that is also highly contextual (relationships, connections, and place)' (Irwin 2015, 238). The Transition Design Framework was introduced at the AIGA National Conference in 2013 by Terry Irwin, Cameron Tonkinwise and Gideon Kossoff who have quickly scaled up the project with publications, symposiums and conversations. The Transition Design approach responds to wicked problems with strategies drawing on sociotechnical transition management theory, the Transition Town Network (see p. 105), the Great Transition Initiative (Kenneth Boulding, Global Scenario Group, the Tellus Institute) and the New Economics Foundation's 'The Great Transition' (Irwin et al. 2015, 2–3). Transition Design relies on concepts and practices such as living systems theory, futuring, indigenous wisdom, cosmopolitan localism, everyday life discourse, post-normal science, needs, social psychological research, social practice theory, alternative economics, worldviews, and Goethean Science and Phenomenology (Ibid., 3–4). It integrates these theories and practices into a framework with four areas of 'knowledge, action and self-reflection: (1) Vision; (2) Theories of Change; (3) Mindset & Posture; (4) New Ways of Designing' (Ibid., 5) (see Figure 1.2). Future-oriented visions inform new projects. The framework emphases that theories of change are always present in a 'planned/designed course of action, whether it is explicitly acknowledged or not' (Ibid., 5). The change that Transition Design seeks is brought about through the acknowledgement that 'fundamental change is often the result of a shift in mindset or worldview that leads to different ways of interacting with others' (Ibid., 6). Transition design is informed by values and knowledge emerging from this new mindset: 'A shift to a more holistic/ecological worldview is one of the most powerful leverage points for transition to sustainable futures' (Ibid., 4). Transition Design is distinct from earlier design practices in its explicit transdisciplinary engagement and 'its understanding of how to initiate and direct change within social and natural systems' (Irwin 2015, 237). It is the most ecologically engaged of all design approaches apart from many types of design activism.

I define **Design Activism** as design work that is explicitly linked to the activism of social movements and/or work with marginalized communities from a perspective of solidarity, allyship and intersectionality. Other theorists have defined 'Design Activism' in different ways. In a book titled *Design Activism*, Alastair Fuad-Luke describes the practice as 'design thinking, imagination and practice applied knowingly or unknowingly to create a counter-narrative aimed a generating and balancing positive social, institutional, environmental and/or economic change' (2009, 27). Since this definition captures many design activities outside of work created in solidarity with social movements and marginalized groups, I consider it far too broad a definition. Not everyone who is interested in using design to address social or environmental problems will engage with the social and political critique of social movements or the distinctive practices and cultures they create and maintain. These networks are spaces for political and social learning that are fundamental for social change. Almost all significant historical example of large-scale social change on issues of oppression is the result of people organizing collectively towards common goals as part of social movements. In a review of the Hopenhagen campaign (see Chapter 12), I describe why whom you are working *for* or *with* matters on issues of social change.

Design activism is design work *with* social movements, mutual aid networks and/or marginalized and sometimes front-line communities. Design activism involves learning from activist cultures and marginalized communities while supporting them with design activities. Some examples of design activism are the various projects of The Center for Urban Pedagogy in Brooklyn (including excellent resources online); of Brave New Alps (including the Precarity Pilot and refugee community economies spaces in the Italian Alps); Occupy Design UK (Noel Douglas and others including myself); and the 'Privileged Participation: Allying with Decoloniality in a Difficult Climate' workshop at Carnegie Mellon School of Design (2016) (run by Dimeji Onafuwa, Jabe Bloom and Teju Cole). Design activists take an allyship approach to social justice work that is significantly different from traditional charitable or *pro bono* work. With this attitude, design activists not only address immediate problems with design skills, but they also learn how to be inclusive in the process of working *with* people facing oppressions and hardships. Due to the structural problems faced by activists, many people with experience doing this work end up rejecting the premise that markets and existing institutions can deliver sustainability. Thus, design activism, reflecting the ethos and practices of activist cultures, networks of solidarity and social movements globally, often works to confront capitalist developments and create post-capitalist alternatives. The anti-oppression, anti-capitalist and post-capitalist analysis developed by social movements informs anti-authoritarian, decolonializing and post-capitalist strategies and practices. For many people in this tradition, the reorientation of design practice towards basic social and ecological values over short-term profit for those with capital remains the most serious design problem that must be addressed.

Notes

1. Within capitalism the ideology of the 'invisible hand' obscures the mechanisms (i.e. the centralized policy decisions that set the rules of the market) that determine whose interests are served by market processes.
2. This is a controversial statement. Most climate scientists and refugees will agree.

2 Design as Symbolic Violence

Design influences beliefs, feelings and sensibilities in subtle ways that are not always obvious to its audiences and users. As a cultural practice, design also functions as a means of negotiating social relations as it constructs the communications, products and spaces that influence how we live. It can work to make new ways of doing things possible by appealing to what people find meaningful, useful and desirable. While it functions in all these ways, design typically reproduces cultural assumptions associated with shared ideas about what constitutes good design and notions of taste. These social norms and conventions can also be challenged and transformed. A good place to start with this work is with the social theory of how power, ideologies and sensibilities are reproduced through cultural processes.

The term 'symbolic violence' was coined by Pierre Bourdieu in *Distinction: A Social Critique of the Judgement of Taste* in 1979 (translated to English in 1984) and later developed in greater depth in subsequent publications such as *Masculine Domination* (2001). While Bourdieu first used the term 'symbolic violence', its conceptual architecture was developed by feminist, race and indigenous scholars and activists over several decades. These groups have described how patriarchy, sexism, racism, colonialism and imperialism exist within oppressive structures and also within cultural practices that embed domineering ideologies into everyday life. Symbolic violence is the result of systems of representation that normalize the hierarchical devaluation of certain people. Symbolic violence enables real violence to take place, often preceding it and later justifying it. It works beneath the level of ideology and becomes a totality of learned habits and sensibilities. It describes a process where controlling concepts are embedded within social practices and ultimately as embodied dispositions. Through symbolic violence, individuals learn to consider unjust conditions as natural and even come to value customs and ideas that are oppressive (towards others and even towards oneself). Symbolic violence also describes the process of the embodiment of domination. It is a form of power that that becomes embedded in 'dispositions deposited, like springs, in the deepest levels of the body' (Bourdieu 2001, 38). It is perpetuated by cultural practices and myths our culture circulates that have 'a hypnotic power' which internalize sensibilities and values systems (Ibid., 9). Since design is a cultural practice involved with the production of things and representations in everyday life, and designers make things and images that purposely evoke particular sensibilities – the theory is significant for design.

Bourdieu's *Distinction* is based on his sociological research that attempted to understand social inequity and why it is that people submit to injustices without resistance. The research examined 'taste' in French society with a survey of 1,217 subjects in the 1960s. Subjects were asked about their tastes in music, art, design, etc. With this research Bourdieu described how class power is reproduced through ideology embedded within cultural and educational practices. Bourdieu describes symbolic violence as: 'a gentle violence, imperceptible and invincible even to its victims, exerted for the most part through the purely symbolic channels of communication and cognition (or more precisely, miscognition), recognition, or even feeling' (2001, 2). The effects are embodied in 'bodies and minds by long collective labour of socialization' (Ibid., 3) and 'exerted not in the pure logic of knowing consciousness but through the schemes of perception' (Ibid., 37). It is through processes of symbolic violence that power imbalances are naturalized: 'The most intolerable conditions of existence can often be perceived as acceptable and even natural' (Ibid., 1). It is because oppressive ideologies are so deeply embedded within policy, custom, language, taste and emotions that issues like sexism, racism and colonialism are so difficult to eradicate.

The Ecocidal Logic of Anti-Environmentalism: Ecoism

Ideologies and practices that are harmful to the environment are at least as difficult to overcome as those that are oppressive to groups of people. The traditions of ecofeminism, social ecology, decolonization and indigeneity critique anthropocentrism as an anti-environmental ideology that presents humans as separate from and more important than everything else. Anthropocentrism elevates humanity as the only species with rights and dismisses the ecological basis of human existence. Since humans are interdependent with their ecological context, it is a dysfunctional worldview. The ecocidal logic of anti-environmentalism can be captured by a term that alludes to the relationship between social and ecological oppressions. 'Ecoism' refers to patterns of thought that enable ecologically destructive behaviour. Ecoism is like feminism, sexism, racism, classism and imperialism – it is a word that responds to a type of oppression. Ecoism defines a way of thinking that denies the value of the ecological context. Like the other 'isms', it is reproduced by cultural practices, social dynamics and social structures through processes of symbolic violence.

Social Theory of Power

Social theory describes how knowledge is socially and historically constructed and bound by power relationships. Knowledge is never neutral. The values, assumptions and 'knowledge' of the powerful are embedded in discourses that are reproduced through cultural processes and established as legitimate. Dominant discourses reflect the points of view of those with the power to monopolize the production and use of knowledge. Discourses can be a form of discipline that control what is recognized as truthful:

> Each society has its regime of truth, its 'general politics' of truth; that is, the types of discourses which it accepts and makes function as true; the mechanisms and instances which enable one to distinguish truth and false sentences, the means by which each is sanctified; the techniques and procedures accorded value in the acquisition of truth; the status of those who are saying what counts as true. (Foucault 1980, 131)

Normative assumptions are influenced by ideology that can appear to be neutral, natural and inevitable. Michael Foucault's concept of 'disciplinary discourse' describes how regimes of truth work without overt coercion to influence belief systems and behaviour: 'When discipline is effective, power operates through persons rather than upon them' (Edwards &

Usher 1994, 92). Herein persons embody ideologies embedded in the discourse, even if these discourses work against their self-interest. Political scientist John Drysek explains that 'discourses can embody power in the way they condition the perception and values of those subject to them, such that some interests are advanced and others are suppressed' (2005, 9). With disciplinary discourse power 'reaches into the very grain of individuals, touches their bodies and inserts itself into their actions and attributes, their discourse, learning, processes, and everyday lives' (Foucault 1980, 39). In communications processes, discourses 'systemically form the objects of which they speak' (Foucault 2002, 54) and so discourses embody and also construct power relations. Both symbolic violence and disciplinary discourse are processes through which cultural ideas are reproduced and internalized into subjectivities.

Stephen Lukes' Theory of Power offers a complementary description of the ways in which power functions in society: where power is most effective when it is least observable. Lukes describes power as consisting of three dimensions:

Lukes' Theory of Power: Three-Dimensional Power

1 One-dimensional power involves the capacity to directly influence events.
2 Two-dimensional power is the ability to influence the agenda of possibilities that may be addressed (and exclude certain perspectives).
3 Three-dimensional power is the ability to control the frameworks through which we make sense of and understand ourselves and our world; this is the most fundamental exercise of power, power used to shape the way we see their world so that we may accept things as taken-for-granted because there appears to be no alternative. (Reason 1998, 150 based on Lukes 2005 [1974])

The third dimension of power is the power exercised through the propagation of powerful ideologies. Power struggles operate across these three dimensions. Since all discourses carry points of view, the most disingenuous are those who present their ideas and the ruling regime as ideologically neutral. The third dimension of power distorts judgement through setting a particular ideological vision as an unquestioned truth. For example, the idea that the market will deliver widespread prosperity and can be used to solve all problems is an ideological proposition. With the third dimension of power, frameworks that people use to understand the world are taken for granted and the ideological is never questioned. The failure to recognize ideologies as assumptions (that might or might not be true) creates severe vulnerabilities (as assumptions remained unexamined). Those who claim that we live in a post-ideological world are attempting to claim their own ideological perspectives as the only option by exerting this third dimension of power (the control of frameworks).

This analysis explains a function of the advertising industry. Lukes asks: 'Is it not the supreme form of power to get another or others to have the desires you want them to have – that is, to secure their compliance by controlling thoughts and desires?' (2005, 27) The role of controlling desire is performed with great skill by the advertising industry with all the knowledge, tools and skills of the social sciences (especially psychology and cognitive science) and the creative arts (design and the media arts). Designers who are aware of their role in reproducing the values supporting conspicuous consumption and the ideological conceits of neoliberalism often maintain the illusion that there are no alternatives. But it is only the widespread belief that there are no alternatives that prevents other options from being explored and demonstrated.

These theories of power clarify why ecological literacy remains marginal over two and a half decades after its introduction, after five decades of vigorous environmental campaigning and facing increasingly serious risks to both non-human nature and civilization. Discourses that are unfavourable to the interests of the powerful are marginalized or even excluded from public debate and the spaces where important decisions are made. Dominant discourses in the public sphere reflect the interests of power. Powerful groups effectively maintain the status quo, that is, unsustainability, through current development regimes, 'rational' economics and a plethora of public relations and advertising programs to reassure the public that business-as-usual is not only morally acceptable but that there is no alternative.[1] With ideology and with system structures, neoliberal capitalism reinforces the belief that design must always serve an economic function first and foremost. From this perspective, destructive ecological and social consequences of the 'normal' operation of the economy are unfortunate, but inevitable. In far too many places, ecologically informed discourses remain marginal. Communication design perpetuates structural injustice through symbolic violence (Bourdieu), disciplinary discourse (Foucault) and the third dimension of power (Lukes). These sociological theories inform the way design reproduces power relations and unsustainable design practice.

Design as Symbolic Violence

Design embeds ideas in communication and artefacts in artful and psychologically powerful ways. The theory of symbolic violence sheds light on how design can reproduce values and priorities, naturalize these ideologically loaded points of view and obfuscate this process. Design functions as symbolic violence when it is implicated (either directly or indirectly) with the creation, facilitation and reproduction of social and ecological harm. In this mode, design conceals both specific ecological and social impacts and wider exploitative dynamics. Through symbolic violence, the design industry functions as an enabling process for the exploitation of certain groups of people and non-human nature. It does this by creating and manipulating tastes, desire and identity to maintain the hegemonic, seductive and enchanting power of an increasingly intensified and authoritarian variety of neoliberal capitalism. Symbolic violence is perhaps most obvious in communication design, although other types of design are also culpable.

Anthropologist Juris Milestone claims that design functions through 'the operation of symbolic violence, as in the ability of dominant groups to establish their sensibilities as the norm, while making this dominance appear to be quite natural and thus non-dominating' (2007a, 86). He describes designers as experts in subjectivity who create 'order by manufacturing certain subjectivities' (2007b, 175). Design disciplines the public by encouraging social hierarchies where people distinguish themselves with their 'good' taste and commitment to quality (Ibid., 178). Design not only drives consumer desire but 'can work to depoliticize war, technology, architecture, consumerism and globalization' (2007a, 96) by virtue of its aesthetic appeal and sophisticated grasp of cultural ideas.

In order to appear ethically oriented, corporations depend on the remoteness of socially and ecologically harmful industrial processes. (In many sectors the conditions under which production processes occur are highly guarded secrets: this is evident in sweetshops as it is in the food industry.) For this reason, a primary role for design is in camouflaging the specific ecological impacts of particular extraction, production and disposal processes and the wider systemic dynamics of human exploitation. Design in this mode conceals the social and ecological unintended consequences of individual products and consumerism. With obscuring and concealing strategies, design cultivates capitalist and neoliberal values and sensibilities enabling exploitation of environmental spaces and structural violence of various types.

Neoliberalism depends on having its assumptions and ideological premises accepted and internalized. Political theorist William Connolly explains, 'neoliberalism must become an ideological machine embedded deeply into life to produce the submission and self constraints its punitive success demands' (2013, 59). Since neoliberalism is dependent on populations that will accept its assumptions[2] (including the dismissal of environmental problems) one of the primary roles for design in this context is to create an illusion of wholesomeness and to whitewash/greenwash[3]/datawash[4] destructive industrial processes. Alternatively, design can be a tool for the illustration and construction of an ecologically viable future. But capitalism needs designers to drive consumption – not critique consumption! Herein lies a basic impasse when attempting to address environmental issues by undesigning consumption within present-day economic 'imperatives'.

Resisting Symbolic Violence by Design

An awareness of the dynamics that perpetuate oppressions, exploitations and ecocide is not enough to end their reproduction. Intellectual engagement and even conscious-raising are insufficient. While the concept of symbolic violence can help reveal injustices, awareness does not necessarily break deep-seated prejudices:

> It is quite illusionary to believe that symbolic violence can be overcome with the weapons of consciousness and will alone, this is because the effect and conditions of its efficacy are durable and deeply embedded in the body in the form of dispositions. (Bourdieu 2001, 39)

While there is no easy solution, it is possible to intervene and transform practices, institutions and system structures that legitimize and reproduce domination in its various forms. This work can be facilitated by design. Those seeking change must simultaneously address the social practices and social structures that perpetuate symbolic violence.

> These critical distinctions are not at all gratuitous. They imply that the symbolic revolution called for by the feminist movement cannot be reduced to a simple conversion of wills. Because the foundation for symbolic violence lies not in mystified consciousness that only need to be enlightened but in dispositions attuned to the structure of domination of which they are the product. (Ibid., 41)

Discovering how symbolic violence works is only the first step towards building capacities to challenge it. Feminist, race, indigenous and decolonialization scholars and activists have been doing this work for decades with struggles against patriarchy, racism and colonialism. Many of the strategies used to confront these oppressions are important touchstones for resisting ecoist[5] forms of symbolic violence. Non-human nature is exploited using the same intellectual strategies of dismissal and othering that have been historically used against women, people of colour (POC) and indigenous people (see Chapter 5). Oppressive frameworks have cast women and POC as 'the other', without agency of their own, with few needs or histories of their own. This discursive marginalization functions to legitimize oppression.

The concept of symbolic violence challenges and ultimately enhances the notion of false consciousness. False consciousness is a Marxist concept that holds that an individual can be 'deceived into complicity with oppression and will therefore unintentionally think and behave in ways that harm herself' (Strega 2005, 221). Critical consciousness is a process of diminishing complicity with oppression, especially forces that work against one's own interests or values (as is the case when one internalizes concepts about one's race, gender and sexuality as inferior). The concept facilitates new agencies by supporting an understanding of 'complicity without pathologizing it or attributing it to underdeveloped consciousness' such that agency becomes a 'capacity to recognize discursive constitutions of self' (Ibid., 222). This analysis of how oppression works takes the pressure off of individuals and locates the problem in the social forces that reproduce ideology. Culture trains us to accept ideas that enable oppressions. Noticing the process is foundational to the development of new capacities for resistance and renewal. Yet acknowledging a role in the reproduction of social and ecological harms is often a profoundly difficult process. Individual consciousness change can be a starting point, but social change must include building alternatives and transforming power relations. This often involves uncomfortable confrontations with those advocating business as usual. Change makers need to be capable of enabling these encounters.

Social movements have demonstrated how breaking oppression is dependent on circumventing, resisting, intervening and ultimately transforming the dynamics that enable oppressive power to reproduce itself. Design can help intervene to break symbolic violence by exposing and reconstructing the processes and practices through which the whole social order is reproduced. One way by which it can do this is by supporting the construction and application of Counterpower.

A Theory of Change: Counterpower

Counterpower is a theory of change that describes how the pressure applied by social movements and especially by the resistance of the oppressed creates social change. Social change theorist and activist Tim Gee explains how ordinary people make change happen:

> When governments, corporations or other ruling institutions yield power, it is not through the goodness of their hearts. It is to save face when the people themselves have already claimed power. Of course, in theory, power can be voluntarily given away by those who already have it, but this has happened only rarely. (2011, 16)

Counterpower is the ability to challenge, remove and/or transform the power of the ruling order. Historically, movements for democracy, expanded suffrage, emancipation, anti-slavery, women's rights, civil rights, human rights, minimum wage and worker's rights have all used counterpower with various degrees of success. They have won extension of voting righting rights, public services, independence from colonial rule, the end of slavery, equal rights for women and people of colour and even revolution from authoritarian states: all rights that those with power were initially not willing to concede. All of these struggles are ongoing.

Counterpower emerges through critical consciousness, based on an understanding of the historical and social context of current social problems as well as the potential for collective action to address these problems. Social control is maintained by ideologies that function to disempower. Counterpower becomes possible once people organize strategically to challenge injustices and the ideologies, institutions and system structures of the ruling order. But awareness alone is not enough: 'personal morality and collective consciousness are both helpful, even necessary for social change. But they are not themselves strategies' (Ibid., 38). The work of transforming structural injustices and ecocidal institutions is the work of organized movements for social change. Gee's theory of change through Counterpower has four stages:

A Theory of Change: Gee's Theory of Counterpower

1 Consciousness is the stage of realizing that there is a problem and creating the conditions for Counterpower.
2 Co-ordination is the stage of building Counterpower through a movement to challenge the problem.
3 Confrontation is the stage when Counterpower is used most intensely, as the movement challenges the target's power outright.
4 Consolidation is about maintaining Counterpower, adjusting to the new balance of power following the Confrontation Stage, and ensuring that it turns into real-life change (Ibid., 130).

This theory of change explains how individuals, movements, campaigns and revolutions make dramatic social change possible. Counterpower works on a micro-level of everyday life when an individual stands against oppression. Individual expressions of Counterpower sometimes trigger social movements when they come to symbolize the injustices embedded in the system.

Counterpower works in the realm of concepts, economics and bodies as idea power, economic power and physical power (Gee 2011, 13). Elites exert their power by persuading, paying and punishing (Ibid., 18). Idea power is perhaps the most insidious in the way it imposes the ideology of oppression on the oppressed. South African anti-apartheid activist Steven Biko famously said: 'The most potent weapon of the oppressor is the mind of the oppressed.' In Western democracies idea and economic power are dominant means of maintaining social control, although certain groups of people are subject to physical violence in various forms. Already the United States locks up more prisoners both per capita and in gross terms than any other major nation, even nations many Americans consider to be authoritarian. In response these three types of power social movements apply three types of counterpower: idea, economic and physical counterpower.

Idea counterpower challenges the ideas, ideologies and legitimacy of the ruling order. Communication design is a powerful means of exerting idea counterpower. It can do this by exposing problems, building counter-narratives and illustrating alternatives. Examples of idea counterpower are radical newspapers, campaigns and teach-ins. The work of designers Tzortzis Rallis and Lazaros Kakoulidis for *The Occupy Times* is an example of design activism in support of idea counterpower. The wider Occupy movement provided memes ('the 99%') and other ideas that transformed public debate on the legitimacy of austerity

Fig. 2.1 *The Occupy Times*, #07. Lazaros Kakoulidis and Tzortzis Rallis, 2011.

in the wake of the banking bailout of 2008. During this time of unprecedented transfer of public wealth to the financial sector and the consequent loss of public services and the start of an era of crippling austerity, Occupy responded with social critique missing in the mainstream media. Activists created news with direst actions, occupations and articulate social critiques. Rallis and Kakoulidis have made the design as radical in style as in content (see Figures 2.1 and 2.2). Jonathan Barnbrook contributed a logo for Occupy London and his Bastard font is used in *The Occupy Times*. In early 2012, Noel Douglas, Tony Credland, Lazaros Kakoulidis, Tzortzis Rallis and others (including myself) founded Occupy Design UK to work with and support the wider Occupy movement.

*Economic counterpowe*r is the 'refusal to work or the refusal to pay. The building of alternative economic power bases – such as trade unions, co-ops, progressive businesses, NGOs and publicly owned services can also be seen as a form of economic Counterpower' (Ibid., 24). Other examples of economic counterpower tactics are boycotts, strikes and disinvestments. A historic example of economic counterpower facilitated by design is the stamp created for the paper *The Poor Man's Guardian* in the early nineteenth century. The stamp bearing the words 'Liberty of the Press' and 'Knowledge is Power' (Figure 2.3) was made in response to a new tax law requiring all papers to have an official tax stamp, which made publishing newspapers prohibitively expensive for the poor and would have effectively defeated the working class publishing and distributing a newspaper. The people who made and distributed *The Poor Man's Guardian* were willing to disobey the law in order to continue publishing. About 740 people were sent to trial for selling newspapers stamped with the unofficial 'Knowledge is Power' seal but the newspaper continued to sell 20,000 a week (Gee 2011, 139) and was foundational to working class social movements. This episode illustrates how hard people have fought historically to have access to news from non-elite sources, as a precursor of *The Occupy Times*.

Fig. 2.2 *The Occupy Times*, #20. Lazaros Kakoulidis and Tzortzis Rallis, 2013.

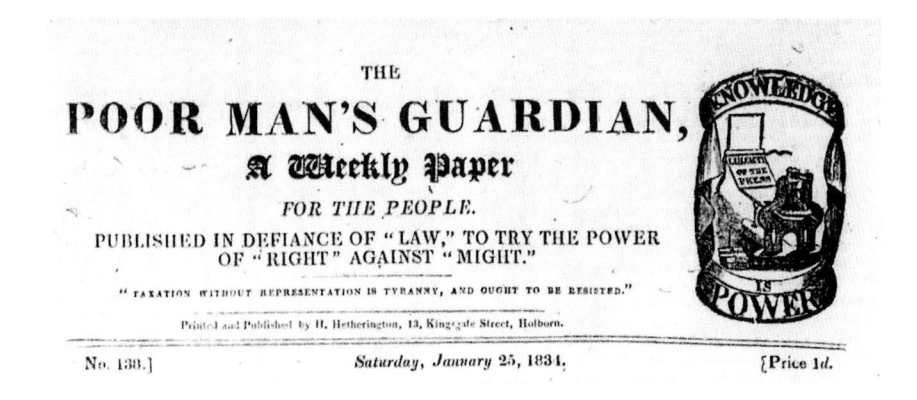

Fig. 2.3 *The Poor Man's Guardian. A Weekly Paper for the People.* Published in defiance of 'Law', to try the power of 'Right' against 'Might'. 'Taxation without representation is tyranny, and ought to be resisted'. Published by Hetherington, 13 Kingsgate Street, Holborn. 25 January 1834.

Physical counterpower is where power is exerted most directly. Non-violent forms of physical counterpower include refusing to cooperate or physically putting one's body in the way with direct action. The government exerts its physical force with bailiffs, the police and the military. Social movements exert physical counterpower with tactics such as blockades and occupations. A historical example of physical counterpower is the Underground Railroad in America. A more recent example facilitated by design are the shields used by Climate Camp activists during the occupation of British Airways at Heathrow Airport in 2007 (protesting the proposed construction of a third runway) (Figures 2.4 and 2.5). The shields featured large images of individuals around the world vulnerable to climate change. The 'face shields' were used to protect the protesters from police batons. Once the activists occupied the British Airways parking lot, the shields were revealed to be cardboard boxes with tents inside. These were erected during the overnight occupation. The shields were featured on the cover of *Creative Review* and exhibited at the V&A's 2014–2015 blockbuster *Disobedient Objects* exhibition. The third runway at Heathrow was not built.

Design can play a role in all four stages (consciousness, co-ordination, confrontation and consolidation) and all three types of counterpower (idea, economic, physical). Activists strategies, techniques and tools are refined, improved and documented in publications and websites such as Andrew Boyd's tactical handbook *Beautiful Trouble: A Toolbox for Revolution* and the publications of the Tactical Technology Collective.[6] These strategies might seem marginal – but social change has always started with small groups of highly devoted people who gain notice through their actions, the authenticity of their commitment and the popular appeal of their message. As they come to be supported by others social movements build organizational capacities to challenge greater injustices. Social movements have transformed societies. While these transformations have often been accomplished non-violently on the part of the activists, activists themselves are routinely subject to state and other forms of violence. The charity Global Witness reported at least 908 documented deaths in thirty-five countries between 2002 and 2013 people killed for their work on

Fig. 2.4 Shields (front and back) from the Climate Camp Heathrow occupation. Featured on the cover of *Creative Review* and in the V&A's *Disobedient Objects* exhibition. PSC Photography, 2007.

Fig. 2.5 Climate Camp shields in use at the Heathrow Camp for Climate Action 2007. Photos by Kristian Buus, ©kbuus, 2009.

environmental and land issues, with only ten convictions (Global Witness 2014, 6). Nearly twice as many environmental activists were killed in 2014 as journalists (Global Witness 2015). Counterpower describes pressure applied by some of the least advantaged, often towards goals that will have universal benefits. Noam Chomsky remarked the 'so-called "least advanced" people are the ones taking the lead in trying to protect all of us' (2013b). When designers work in solidarity with social movements, they are working as design activists (p. 26). Design skills can help build awareness, inform and build solutions – and also protect activists and other people.

Notes

1. UK Prime Minister Margaret Thatcher coined the phrase 'There is no alternative' in the 1980s. This idea has become known by its acronym as the 'TINA Principle'.
2. Two dominant assumptions of neoliberalism are: (1) that markets are the best means to facilitate human relations; and (2) that infinite quantitative economic growth is possible and desirable.
3. Greenwash is the presentation of environmental destructive products and activities as sustainable.
4. Datawash occurs with data visualization practices where the most essential facts are obscured.
5. Ecoism/ecoist is a neologism that links ecological denial to other types of othering and oppressions.
6. Available online: http://beautifultrouble.org and https://tacticaltech.org

3 Design versus The Design Industry

Despite all the creativity, knowledge and emergent skills at our disposal, designers continue to contribute to developments that exacerbate ecological harms. While design continually evolves as a practice with increasing potential for innovation, designers struggle to effectively address eco-social problems in the current context. While there is intense controversy about the direction of various global trends, the fact that climate change, biodiversity loss and other environmental harms are civilization-threatening problems are accelerating is unequivocal. More controversially, inequality is also rising globally (Piketty 2014; OECD 2015; Credit Suisse 2016; Oxfam 2017; see pp. 162–164) and raising inequality is bad for everyone, including the well-off (Wilkinson & Pickett 2010). Eco-social problems can be addressed by design – but this is still not happening on scale. One way to understand why this is not happening is with analysis of the systemic priorities in the design industry. Design must be understood as a broader practice than the work that is produced within the 'design industry'. This distinction is important because the design industry operates according to different logic than the logic that inspires many individual designers. Where 'design' is understood as a socially beneficial activity engaged with building a better world, it is distinct from much of the work created for 'the design industry' with its practices stamped with the assumptions of profit-first market systems, including the norms and conventions that reproduce unsustainability.

Design ≠ The Design Industry

Many designers are motivated to use their skills to address social and environmental problems they see proliferating around them. The design industry, on the other hand, operates according to the logic of capitalism, that is, oriented towards the creation of economic profit (for those with the capital to invest) and quantitative economic growth. The economic system functions without regard to biophysical reality (until resources scarcities, illness, extreme weather, etc. impact the economy). Herein lies a basic impasse: the design industry operates according to reductive feedback based on priorities determined by the capitalist economic system – as opposed to feedback from the systems in which the economic system is situated and upon which it is dependent (the social and ecological systems). Since many individual designers do attempt to address problems outside the scope of market priorities, the practice of design is not restricted to what is economically advantageous. Where *individual* designers recognize the larger context and respond to problems outside of the economic order, the design *industry* is oriented towards the two standard business priorities: capital accumulation and market growth. With this reductive focus, human and natural 'resources' are used to create profit. The needs of the social and ecological orders are systemically ignored, as much as possible at least until crisis conditions make this ignoring impossible. The feedback from the market is impersonal and simple, but the impacts upon the Earth and the well-being of the vast majority of its inhabitants are much more complex.

Designers are financially rewarded for certain tasks. Designers' activities are oriented towards the design of products, communications and buildings that are profitable. The priorities of the design industry are determined by the powerful dynamics of the economic system. In this context, designers often struggle to effectively address environmental and social problems (since solutions over the long term are typically not aligned with what makes short-term profits possible). Perversely, exploiting people and planet is often more profitable than not doing so, and wealth accumulates around actors who take advantage of this dynamic. Problems resulting from this dilemma are intensified by distortions of knowledge and reason due to the influence of the market and anti-environmental ideologies in the media, in political processes (i.e. the climate denial industry, corporate lobbyists, etc.) and even in academia (where, in some places, priorities are set according to what is likely to be profitable research). For designers with social and environmental concerns, embedding these values and priorities within their workplace can be a challenge.

While describing the design industry as oriented towards profit and economic growth is a simplification of its dynamic, the dominant characteristic of business in capitalism is clearly the pursuit of profit. The common use phrase 'the bottom line' is an everyday acknowledgement of this priority. Design firms and individual designers that ignore this imperative struggle to survive. The reductive focus on profit as the highest priority conflicts with the priorities and complexities of the ecological context in which the economic system is embedded. Since the ecological system is the context of capitalism, the ultimate imperative is to sustain this ecological context on which we depend. Occasionally, firms can resist the pressure to exploit people and planet by appealing to a socially conscious target market, but the greater dynamics of the entire economic system is accelerated ecological harm (as is evidenced by grave risks associated with climate change and biodiversity loss amongst other issues).

Ecological and social values struggle to compete in the capitalist economy. As priorities they are systemically devalued as market mechanisms reward those who exploit ecological and human resources for the least investment. For example, those who value the preservation of forests can donate money to charities, but in a market-dominated economy these charities are so marginalized they are not able to stop the rapid destruction of old growth forests and the people and species that depend on these spaces. Corporate communication channels organize flows of information to suit the priorities of industry, which typically deny the severity of ecological harms. Communication, media and design are all oriented towards cultivating values that will support market growth.

Despite these obscuring dynamics, even avid champions of technology have begun to notice the extreme dangers of this fixation with growth. Media, technology and cultural theorist Douglas Rushkoff's recent book is subtitled: *How Growth Became the Enemy of Prosperity*. Here Rushkoff describes how:

> We are caught in a growth trap … we are running an extractive, growth-driven economic system that has reached the limits of its ability to serve anyone, rich or

poor, human or corporate. Moreover, we're running on supercomputers and digital networks that accelerate and amplify all its effects. Growth is the single, uncontested, core command of the digital economy . . . Classical economists and business experts have been of little help. This is because they tend to accept growth-based economy as a preexisting condition of nature. It is not. The rules of our economy were invented by particular human beings, at particular moments in history, with particular goals and agendas. By refusing to acknowledge the existence of this man-made landscape and our complicity in perpetuating it, we render ourselves incapable of getting beneath its surface. We end up transacting and living at the mercy of a system. (2016, 4–5)

In order to address the problems this addiction with growth creates Rushkoff counsels 'we must instead look at the underlying assumptions of the marketplace' (Ibid., 5). Thinking like a programmer (or a designer) is helpful: 'Perhaps ironically, only by thinking like programmers can we adapt the economy to serve human beings instead of the quite arbitrary but deeply embedded ideal of growth . . .' (Ibid., 5). Both design and programming are practices that construct new systems that serve a particular purpose. Good theorists and practitioners in each profession understand how agendas and functions are designed into systems. Both professions are well placed to see the danger of an economic system that has been designed to grow in ways that are destructive to its ecological and social context.

An exclusive focus on profit and quantitative growth in the economic system does not reflect the complexity of systemic conditions. The design industry relies on profit as feedback to establish value. But profits do not reflect ecological stability, resilience, equity, well-being or happiness for the vast majority. Profits generally come from serving the needs of those with the ability to pay – while procuring as many ecological and social resources as cheaply as possible. The narrow focus on economic profit excludes a holistic appraisal of values and encourages short-term thinking and waste of ecological and human 'resources'. Even our language becomes distorted: we know that neither nature nor people are inherently 'resources' but have value in their own right outside of their utilitarian function as a source of profit. This exclusive attention to profit creates three dominant problems:

1. it weakens all other values by making profitability the dominant priority;
2. it rewards those who ascribe to this way of thinking, that is, those who will discount other values; and
3. it's discursive discipline distorts knowledge in the interests of industry and financial capital.

Combined these three dynamics create grave problems, especially for the environment. As the market grows, it consumes everything to suit its own needs: our language, our values and our ideas about what can and cannot be an economic transaction. In a hyper-globalized economy, transnational capital swallows everything in its wake until there is nothing left to use. Evidence of this gorging takes the form of lost species, destroyed rainforests and an unstable climate system: complex ecological systems and species that have evolved over millions of years that are being degraded, destroyed, destabilized and/or made extinct in a matter of a few decades.

The Myth of the 'Free' Market

The ecological system existed before and will be here long after the economic system. Despite this fact, the current economic system was not designed to respond to the needs of the ecological system. Clearly, capitalism was not designed in a studio but evolved over time based on a particular kind of economic theory. This economic theory reflected philosophical and political assumptions of the early (privileged) economic theorists. Consequently, it was designed to serve their interests. Economic decisions over the past two centuries have been based on a certain type of economic theory; that of market liberalism, that is, the belief that supposedly self-regulating markets are the best means of organizing an economy. In 1944, Karl Polanyi exposed the myth of the free market (Stiglitz 2001, xiii) by describing how laissez-faire economics was planned and designed to work in particular ways. Far from being a natural state of affairs, laissez-fair free markets require 'statecraft and repression to impose the logic of the market and its attendant risks on ordinary people' (Block 2001, xxvii). Polanyi wrote, 'There was nothing natural about laissez-faire; free markets could never have come into being by merely allowing things to take their course' (1944, 145). The very notion of 'free markets' is misleading because markets currently do not exist without laws defining who owns what and regulation to facilitate market processes. These rules determine how the economic system functions – and who benefits.

Polanyi's description of the disembedded economy is a key contribution to social thought. Before the advent of market liberalism, described in Adam Smith's the *Wealth of Nations* in 1776, the economic order was always a mere function of the social order (Polanyi 1944, 74). Market liberalism was the first economic system in history to subordinate both the social and ecological systems to the market. Polanyi describes how market liberalism was created with no regard for the social and ecological context in which it is situated. As such, this economic system dangerously ignores the fact that it is embedded and entirely dependent on its social and ecological context (see Figure 3.1). The problems in the social and ecological orders are not recognized until inevitable, increasingly severe crises occur. Ecological theory describes the stable constellations of three domains (economic, social and ecological) where the economic system is designed to function as a subsystem of the social and ecological orders. In a sustainable economy, the market must respond to feedback from its ecological context to enable regenerative development. Instead, the current economic system prioritizes the needs of the market over those of the context in which the market exists. The unstable and stable constellation of domains is illustrated below (Figure 3.2).

Green and ecological economists describe how a narrow commitment to quantitative growth over all other types of feedback undermines opportunities for long-term prosperity. This argument is no longer a radical green idea to those who are paying attention. Mechanical engineer Professor Roderick Smith described the consequences of the fixation with quantitative economic growth in a noteworthy speech at the UK Royal Academy of Engineering:

> relatively modest annual percentage growth rates lead to surprisingly short doubling times. Thus, a 3% growth rate, which is typical of the rate of a developed economy, leads to a doubling time of just over 23 years. The 10% rates of rapidly developing economies double the size of the economy in just under 7 years. These figures come as a surprise to many people, but the real surprise is that each successive doubling period

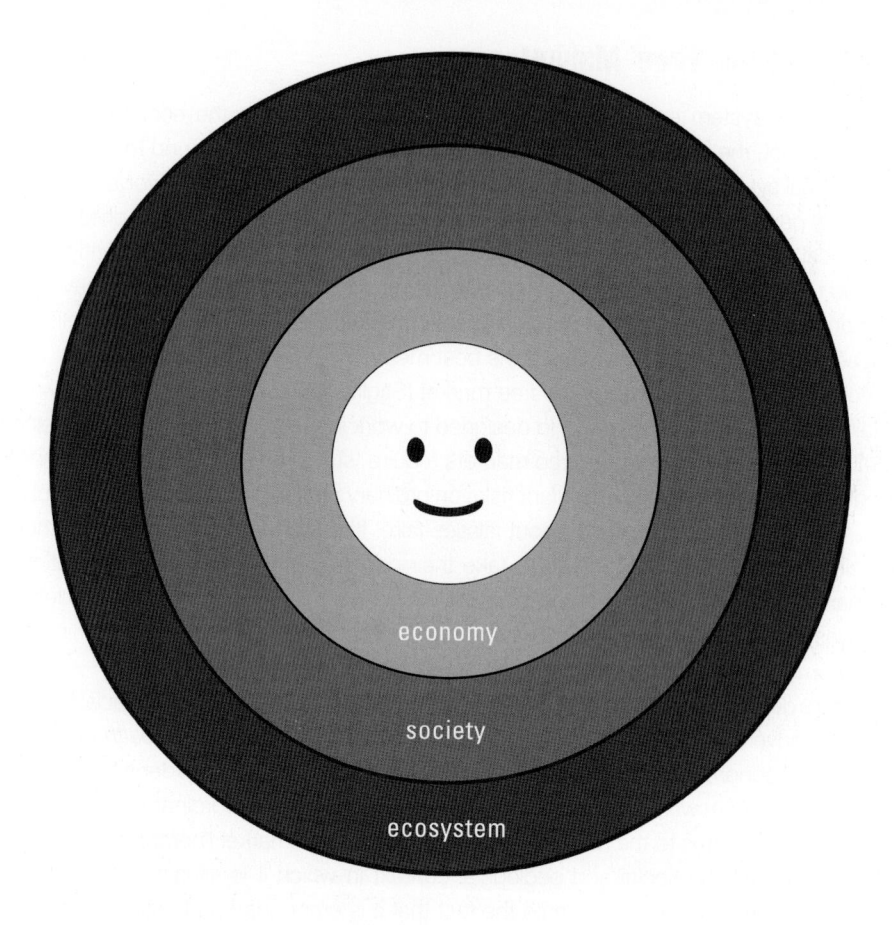

Fig. 3.1 *The Embedded Economy.* EcoLabs, 2015.

The unstable constellations of three domains
CURRENT ECONOMIC SITUATION

The stable constellations of three domains
SUSTAINABLE ALIGNMENT / HIERARCHY OF DOMAINS

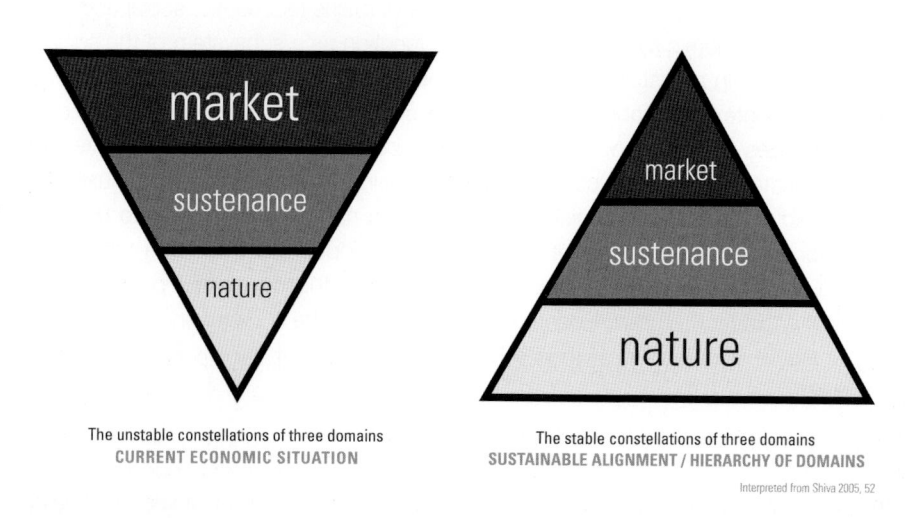

Interpreted from Shiva 2005, 52

Fig. 3.2 *The Stable/Unstable Constellations of the Three Domains.* EcoLabs, 2015 – following Shiva 2005.

consumes as much resource as all the previous doubling periods combined. This little appreciated fact lies at the heart of why our current economic model is unsustainable. (2007, 17)

The expansive dynamics of the capitalist economic system locked into infinite quantitative growth are fundamentally in conflict with the ecological system with (relatively) finite ecological resources and planetary boundaries. This is what James O'Connor described as the 'second contradiction of capitalism' (1998, 165) in which capitalist processes 'self-destruct by impairing or destroying rather than reproducing the conditions necessary to their own reproduction' (Ibid., 13). Since humanity's collective ecological footprint exceeds the Earth's biocapacity (the area available to produce renewable resources and absorb greenhouse gases) by well over 50 per cent, this self-destruction is happening at an alarming rate. Quantitative growth requires a constant increase in flow of ecological resources that results with an increasing flow of pollution (including greenhouse gases in the atmosphere). The Earth has significant buffers and is resilient to quite a lot of abuse but tipping points on various scales are already being triggered with irreversible damage. Shrinking the biocapacity on which we depend is not something that can be sustained indefinitely. Scientists warn of the extreme danger of this situation but the design disciplines have been slow to respond and often completely disengaged. The contradictions in the capitalist growth economy undermine conditions of its own reproduction.

In 2009, the UK Sustainable Development Commission (SDC) published *Prosperity Without Growth?* This report analysed how quantitative market growth threatens not only social well-being and ecological sustainability but also economic prosperity. Author Tim Jackson maintains that neither decoupling nor technological fixes can deliver sustainability in an economy dedicated to quantitative growth. Decoupling economic growth from environmental impacts is a primary strategy of those claim environmental problems can be effectively addressed within the current economic framework. But absolute decoupling of aggregated impacts in the current context is unlikely:

> In a [future] world of nine billion people, all aspiring to a level of income commensurate with 2% growth on the average EU income today, carbon intensities (for example) would have to fall on average by over 11% per year to stabilise the climate, 16 times faster than it has done since 1990. By 2050, the global carbon intensity would need to be only six grams per dollar of output, almost 130 times lower than it is today ... In this context simplistic assumptions that capitalism's propensity for efficiency will allow us to stablise the climate and protect against resource scarcity are nothing short of delusional. Those who promote decoupling as an escape route from the dilemma of growth need to take a closer look at the historical evidence – and at the basic arithmetic of growth. (Jackson 2009, 8)

Ecological economists call for quantitative degrowth (in harmful development) and qualitative growth (in shared prosperity). The political and economic systems must be redesigned to move from a pursuit of infinite *quantitative* growth to a commitment to infinite *qualitative* growth wherein there is dramatic 'degrowth' in harmful ecological impacts. The shift from quantitative to qualitative growth and degrowth in many areas[1] is a radical proposal for a regenerative economy. Ecological economist Herman Daly points out that the first literal dictionary definition of 'growth' is 'to spring up and develop to maturity ... thus the very notion of growth includes some concept of maturity or sufficiency, beyond which point physical accumulation gives way to physical maintenance' (Daly quoted in Simms, Johnson & Chowla 2010, 4). At maturity growth must give way to a state of dynamic equilibrium. Dynamic equilibrium used within this context refers to an economic system that exists 'within ecosystem limits but where there is constant change, shifting balances and evolution' (Ibid., 121). The economy must permit 'qualitative development but not aggregate quantitative growth' (Daly 2008, 1). These ideas are still fiercely contested. A new UK government terminated the Sustainable Development Commission in 2011. Capitalism is structurally committed to quantitative economic growth and governments continues to use the discredited concept of GDP to measure economic growth.

Gross domestic product (GDP) was never intended be used in such a simplistic fashion as it is today. Simon Kuznets, the creator of the GNP/GDP metric, 'warned in 1934 that such a limited, one-dimensional metric should not be used as an index of overall social progress' (Simms, Johnson & Chowla 2010, 4). Since only 1 per cent of growth contributes to rising standards of living (Wijkman & Rockstrom 2011, 128), GDP growth does not equal greater well-being or prosperity for the majority. Fritjof Capra and Hazel Henderson's report *Qualitative Growth* for the Institute of Chartered Accountants in England & Wales describes how growth changes in living systems as they mature. They propose a new concept of quality within market growth:

> Instead of assessing the state of the economy in terms of the crude quantitative measure of GDP, we need to distinguish between 'good' growth and 'bad' growth and then increase the former at the expense of the latter … From the ecological point of view, the distinction between 'good' and 'bad' economic growth is obvious. Bad growth is growth of production processes and services which externalize social and environmental costs, that are based on fossil fuels, involve toxic substances, deplete our natural resources, and degrade the Earth's ecosystems. Good growth is growth of more efficient production processes and services which fully internalize costs that involve renewable energies, zero emissions, continual recycling of natural resources, and restoration of the Earth's ecosystems. (2009, 9)

The difference between 'good' and 'bad' growth from a social and an ecological perspective is clear. Bad growth increases inequities and demands an ever-increasing flow of energy and natural resources extracted from the Earth (often with damaging local consequences) moving through the economic system and generating toxic wastes and greenhouse gases. Good growth increases equity, commonality, democracy and peaceful relations. Good growth can also be regenerative and increase biocapacity. It uses energy directly from the sun, leaves no pollution and increases socio-ecological resilience.

The degrowth movement challenges to the hegemony of the growth paradigm. It calls for a new economics with a democratic and redistributive downscaling of production and consumption in rich nations. Degrowth scholars describe how the logic of growth and accumulation leads to various types of expansions: of new territories (land-grabbing, privatizations, monetarization of ecosystem services, etc.); of developments in risky areas (fracking, geo-engineering, GMOs, deep-sea drilling, etc.); and in the capacity to exploit humans and natural 'resources' (Muraca 2016). The growth economy also leads an intensification of pace of life (via competition and accompanied pressure and anxiety) with increasing debts and increasing commodification (Ibid.). The growth economy is based on two myths: (1) trickle-down economics and (2) economic growth as leading to an increase in gross happiness. But social inequity is not addressed by growth as it ultimately relies on exploitation of people and resources in a globalized context (Spash 2016). The Degrowth movement has been gaining traction (especially in Europe). The idea is even mentioned in *Encyclical Letter Laudato Si*. By Pope Francis:

> In any event, if in some cases sustainable development were to involve new forms of growth, then in other cases, given the insatiable and irresponsible growth produced over many decades, we need also to think of containing growth by setting some reasonable limits and even retracing our steps before it is too late. We know how unsustainable is the behaviour of those who constantly consume and destroy, while others are not yet able to live in a way worthy of their human dignity. That is why the time has come to accept decreased growth in some parts of the world, in order to provide resources for other places to experience healthy growth. (2015, para 193)

While degrowthers are widely condemned by traditional economists as utopian, they counter that there is nothing more removed from reality than an economic system that denies the geophysical context (D'Alisa, Demaria & Kallis 2014). Degrowth offers an alternative vision based on the Earth sciences, for regenerative economic development in industrialized nations.

Distortions of Knowledge and Reason

Knowledge suffers when market dynamics determine what is valid and reasonable. In response to growing public concern on climate issues Exxon, the Koch brother and other fossil fuel interests funnel millions of dollars into deliberate misinformation campaigns, climate denial groups and contrarian lobbyists to misrepresent climate science (Oreskes 2010; Brulle 2014). Funding of science museums by fossil fuel companies not only undermines the knowledge dissemination role performed by these institutions but functions to legitimize the industry most responsible for climate change. Other examples are less obvious but equally serious. Within cultural institutions, education and the media environmental concerns are sidelined (see Chapter 2). In the United States, even basic scientific information on climate change or evolution is negligible in places – including the United States of America's House of Representatives Committee on Science, Space and Technology. Design can amplify these problems – or it can help reveal these obfuscations.

Sustainability requires decreasing resource use, but decreasing consumption threatens the profitability of industry. Thus industry and some government bodies often avoid solutions that involve less consumption (with a few well-publicized but often token exceptions). Corporations suppress information that threatens their own capacity to increase profits – sometimes actively (with misleading advertising, public relations, corporate lobbying and the climate denial industry) but more often passively (such as creating conditions where honest communication of environmental problems and solutions is marginalized to the point of obscurity). Designers are employed to facilitate these processes.

Design skills are harnessed for commercial imperatives. Creative directors working for global brands are rewarded with large salaries (although the agency workers are not so fortunate). When designers attempt to address social and ecological problems, problems outside work dictated by the market, problems with no clear consumer or client, they often struggle to pay the bills. In addressing social and ecological problems, designers expose themselves to financial ruin (as few commercial clients will pay for unprofitable work of protecting the environment). The space to do this work becomes increasingly precarious as wealth is concentrated with new corporate enclosures on common resources (i.e. the privatization of public institutions and the ecological commons). The failure of the design industry to reflect priorities associated with preserving the planet and creating healthy communities creates stark choices for individual designers who want to address systemic problems but are often forced to earn a living by perpetuating destructive market processes.

Environmental communication is one of the areas where the distortion of knowledge is most dramatic. For example, communication generated by environmental organizations is minuscule in comparison to that produced by corporate communication. The advertising industry creates very different representation of the capacity of the natural world to tolerate industrial exploitation. Industry has plentiful resources to communicate a view of nature that suits its own needs. For example, advertising across all media was expected to reach $563 billion by 2017 (Austin, Barnard & Hutcheon 2016, 9). In 2014, the advertising industry expenditures were: UK $22.5bn USD; USA $176bn USD and worldwide $519 billion USD (Austin, Barnard & Hutcheon 2014, 11–10) [equalling approximately 1 per cent of respective GDPs]. The advertising industry uses the talents of designers to illustrate the green credentials of their own products and brands while also attempting to reassure audiences that business-as-usual is morally sound. The visibility of advertising marginalizes environmental concerns to the point of obscurity while creating a characterization of nature as infinitely exploitable.

In sharp contrast to the money available for corporate advertising, environmental organizations have miniscule budgets to communicate responsible information on the state of the environment. For example, the combined entire annual budget (income) for WWF-UK (£63.2m), Greenpeace UK (£5.3m) and FOE_UK (£10.2m) is approximately £78.7m ($98m USD) (Caritas Data 2016). While NGOs are able to leverage their causes due to the gravitas of their mission and thus in some media environmental discourses are visible without the support of environmental NGO sponsorship, this visibility is higher only in some types of media. This medium does not have the scope of commercial advertising. It reaches only those who read environmental news or watch environmental documentaries.

Corporate advertising is a primary means of sense-making in this market-dominated society. Advertising creates representations of nature that suit its own purposes, that is, typically promoting continued conspicuous consumption and presenting nature as open to continued abuse. Herein lies a basic failure in communication systems due to the hegemony of the market as a system that influences the production and reproduction of 'knowledge'. As corporate media takes over traditional scientific publications such as *National Geographic* (the 127-year old magazine was purchased by Robert Murdoch's 21st Century Fox in 2015), environmental communication is even more directly associated with a particular ideological agenda. The result is distortion of knowledge in the public's perception of the conditions of the natural world and associated risks to civilization. While communication channels such as social media offer a means to contest corporate messaging, the dominant channels constantly reinforce the underlying assumption that business-as-usual can continue into perpetuity. Since accurate information is a basic foundation for designing effective responses, the distortion of knowledge on issues concerning the environment makes dealing with already-difficult problems even more intractable.

Environmental Communication as a Crisis Discipline

Environmental communication is a field that is recognized as beset by crisis. In a seminal essay, communication theorist Robert Cox described environmental communication as a 'crisis discipline' dealing with issues of urgency, reflecting the rapid change documented by the Earth sciences and conservation biology (2007). For this reason, Cox argues that there is 'a principal ethical duty of environmental communication: the obligation to enhance the ability of society to respond appropriately to environmental signals relevant to the well-being of both human communities and natural biological systems' (Ibid., 5). Defining environmental communication as a crisis discipline helps establish new premises to serve as guides to strategic action. Following Cox's logic, engaged communication designers can adopt at least four new normative tenets on issues of the environment:

1 Representations of the environment must be transparent, accessible with sources cited.
2 Opportunities to learn from ecology and ecological theory must be nurtured in communication design.
3 Communication designers have a duty to use their skills to support sustainable transitions.
4 Because power often obscures communication on the environment, critical approaches are necessary.

Communicators can help by calling on the attention and resources of other sectors of society to respond. Communication is key to mobilizing responses at moments of contingency and communication designers are central to this enterprise. Design facilitates the development of solutions. Together these practices (communication + design) are pivotal to sustainable transitions.

Socially Responsive Design in Context

While the idea that design is involved with creating a better world is the dominant rhetoric in the design industry and reflects the stated intentions of many designers – designers also simultaneously have other, often-conflicting intentions, reflecting the systemic priorities of the design industry. Despite the aspirational goals of many designers, most design projects start with the identification of a potential for profit-making by servicing the desires of those with expendable capital. Meanwhile, the design industry and the entire capitalist system depends on the ecological system for stability, raw materials and productive capacities; society for stable markets; and on people for labour. Despite these basic facts, the current economic regime systemically ignores the ecological and social orders that provide the context for wealth creation. Market valuation processes reflect neither the ecological nor social 'costs', that is, the so-called 'externalities' of products and services. Herein is a dangerous tension between the economic system and the socio-ecological systems, the design industry and the individual designer. The design industry is situated at this interface between the profit-seeking market and an emerging community of practitioners progressively capable of designing sustainability into the system (in theory) but unable to do so (in practice). While many individual designers are increasingly concerned about our ecological context, they struggle to materialize social and environmental priorities within a capitalism system predominately oblivious to these issues. There are models available for a transition. Goals on this level will require profound shifts in governance systems and corporate culture that could be facilitated by design – if design is liberated from current market priorities. For these reasons, the design of the political economy itself is the primary design problem. Redirected design must help transform the system that determines what is designed.

Note

1. Degrowth is never considered to be a viable universal strategy. Geographic regions with high levels of poverty obviously need growth to improve living standards. Degrowth is about types and distribution patterns of economic growth.

Part 2 Ecology

4 Ecological Theory 101

Ecological thought challenges the tradition wherein nature is regarded as simply resources for humankind to exploit. It proposes that both ecological and social crises result from a legacy of thought wherein the environment is objectified as alien and reduced to chattel for human consumption. The historical processes that have led to the denial of human integration within non-human nature have left a dysfunctional legacy. The modernist descriptions of reality, both scientific positivism and post-modernism, form the theoretical building blocks of the dominant scientific, political and cultural institutions and are a basis of the cultural fabric and the social order. Ecological thought rejects these worldviews insofar as both neglect ecological context. The tradition of reductionism, mechanism, anthropocentrism and radical individualism that characterizes the modernist and the ultra-modernism of much 'post-modernity' must give way to a more integrative way of knowing as a foundation for addressing hyper-complex contemporary problems. In this chapter, I present a brief history of ecological thought as an introduction to seven chapters on how ecological theory informs design.[1]

The scientific revolution (circa sixteenth to seventeenth centuries) saw a complete change in epistemology[2] and ontology[3] from earlier periods replacing the geocentric view of ecclesiastical authority with Nicolaus Copernicus' (1473–1543) heliocentric theory of the movement of planets, Galileo Galilei (1564–1642) theory of gravitation, Frances Bacon's (1561–1626) empiricism, Rene Descartes' (1596–1650) rationalism and Isaac Newton's (1643–1727) mechanism. The result was a new perception and conception of the world: a vision of nature as 'a mechanism whose elements can be disassembled and then put back together again' (Santos 2007, 17). This worldview had 'an ontology that emphasized a mechanistic cosmology, which was primarily determinist, and materialist; and an epistemology that was objectivist, positivist, reductive and dualist' (Sterling 2003, 143). Enlightenment science holds that valid knowledge is derived from empirical evidence consisting of logical, often mathematical and quantitative approaches to sensory experience. Ecological theorists argue the processes through which this study occurs (i.e. the scientific method) create a stark split between subject and object, sensing and thinking, mind and body, humankind and nature. The result is an objectified vision of the universe as mechanical and inert. Positivist science set the stage for the unfolding of industrial exploitation of non-human nature by creating the conceptual framework on which modernity is built. The fact that this worldview serves to justify the exploitation of women and people of colour is not coincidental.

From an ecologically informed perspective, the philosophical underpinnings of the scientific revolution are ridden with methodological and ethical problems, ontological difficulties and even rational contradictions. The father of the scientific method, Frances Bacon, used metaphors that treat 'nature as a female to be tortured through mechanical inventions' (Merchant 2001, 277). In his own words: science would make humanity 'the master and owner of nature' (Bacon 1933 quoted in Santos 2007, 17). Philosopher, ecofeminist and historian of science, Carolyn Merchant claims that Bacon created a powerful cultural metaphor of nature as feminine and as a force that needed to be controlled. Nature is to be mastered and made to submit and to 'take orders from Man and work under his authority' (Bacon quoted in Harding 2006, 26). This approach produced a dramatically new conception of human-nature relations:

> The removal of the animistic, organic assumptions about the cosmos constituted the death of nature – the most far-reaching effect of the scientific revolution. Because nature was now viewed as a system of dead, inert, particles moved by external, rather than inherent forces, the mechanical framework itself could legitimize the manipulation of nature. (Merchant 2001, 281)

Bacon embedded oppressive sexual politics into the tombs of enlightenment science by referring to nature as female and describing science as a means to put nature into 'constraints', to be made a 'slave' and to be 'bound into service' (Bacon quoted in Merchant 2001, 277). Scientific methods should not just 'exert gentle guidance over nature's course; they have the power to conquer and subdue her, to shake her to the foundations' (Bacon quoted in Merchant 2001, 279). Bacon's misogyny and pursuit of domination was embedded into the scientific method and then reflected in the new period's attitude towards the non-human natural world.

With this new perspective nature was devitalized and reduced to various resources to be exploited. Descartes described the natural world as 'a giant machine' that functioned according to mechanical principles and as 'devoid of soul' (Descartes quoted in Harding 2006, 26–27). The 'externalization and de-animating' as well as the 'disenchanting and instrumentalizing' of nature (Fisher 2012, 86, 88) created a new conception of the world as an inanimate machine and removed ethical restraints against exploitation (Ibid., 86). Native American activist Russell Means maintains Western thinkers reduced the complexity of the world to abstractions 'and in doing so they made Europe more able and ready to act as an expansionist culture' (1980). The scientific method condenses the complexity of nature by transforming it into measurable data with the assumption that studying the parts is key to understanding the whole. The regenerative, qualitative, life-creating capacities of the Earth are taken for granted. The machine metaphor led to quantification biases and illusions of absolute prediction and control. Modernity's conception of the world is largely based on these presuppositions.

In the early twentieth century, physicists Einstein, Heisenberg and others made discoveries that recognized the observer as a participant within a process of knowledge-making and demonstrated the problems with subject/object dualism in the scientific method. Quantum physics profoundly challenged thinking on observation and perception, participation, relationships and influences. Shortly thereafter philosophers John Dewey, Charles Pierce and others developed philosophical critiques of Cartesian science. The twentieth century saw increasing criticism of positivism culminating perhaps with the Vandana Shiva's paper 'Reductionist Science as Epistemological Violence'. Here Shiva criticizes positivist science for its arrogance: 'Reductionist science is also at the root of the growing ecological crisis, because it entails a transformation of nature such that the processes, regularities and regenerative capacity of nature are destroyed' (1988, unpaginated). Ecological theorist Boaventura de Sousa Santos claims that this approach to science erases complexity such that 'knowledge gains in rigour what it loses in richness' (2007, 27). Over the past century ecological theory and holistic science has described how 'knowledge gained from observation of the parts is necessarily distorted' (Ibid., 28). The multifaceted critique of positivist science created the foundation for the development of more inclusive and holistic approaches to knowledge enabling ecological thought to emerge as a philosophy supporting mutually beneficial relations with the non-human natural world.

The Problematic Emergence of Ecological Thought

The term 'ecology' derives 'eco' from the Greek οἶκος (meaning 'house') and 'logo' λογία (meaning a 'body of knowledge'). Together, ecology means 'the science of habitat'. The word was coined by Ernest Haeckel in 1866 in *Generelle Morphologie*. Haeckel described ecology as 'the science of relations between an organism and the surrounding outer world' (Haeckel 1866 quoted in Olsen 1999, 68 and Goodbun 2011, 257). Haeckel was also 'an early (holistic-organic) systems thinker in biology, and with Jakob von Uexküll developed the concept of an *environment*' (Goodbun 2011, 46). As well as a being a prominent German biologist and philosopher, Haeckel was a talented artist who created intricately detailed drawings of embryos, plants and microscopic life forms (see Figure 4.1). Haeckel's artwork linked the study of ecology with images from its conception (the links between aesthetics, visuality and ecology will be explored throughout this book). Haeckel gave ecology a creative genesis but also one ridden with concepts that can be easily used in oppressive ways. These ideas continue to create severe problems.

It was Haeckel's ideological inclinations associated to his findings in the natural world that began a troubled legacy between ecological thought and socially deterministic and even racist thought. Haeckel's adoption of Social Darwinian positions with nationalist,

> right-wing *volkisch* philosophy combining vernacular holist beliefs with an ideologically distorted science ... Needless to say, Haeckel's version of organicism [from the concept of 'organic'] proved all too useful to fascist ideologues and this particular political legacy explains the uncomfortable reaction of many contemporary academics to the use of the word [organic]. (Goodbun 2011, 46–47)

Sixty years later philosopher and politician Jan Smuts coined the term 'holism' in the 1926 book *Holism and Evolution*. Smuts, a supporter of racial segregation in South Africa, used ecological concepts to justify racist political agendas.[4] Later the Nazi party in Germany used references to nature to promote the Aryan race as the master race, with the extermination of the '*Untermensch*' ('sub-humans'). These episodes illustrate how ecology as a discipline and associated nature-based philosophies can be used as a means of claiming a privileged perspective on the 'natural' for despotic political agendas.[5] These problems must be confronted with robust anti-oppression theory and practices.

The holocaust made the horror of imperialist and racist nature-based philosophies of the era even more blatantly apparent than previous types of state sanctioned violence. The study of ecology after this period became restricted to positivist, empirically testable methodologies (Sachs 2010, 30). These methods led to the

Fig. 4.1 *Phaeodari. Kunstformen der Natur*, Plate 61, Ernest Haeckel, 1904.

development of ecological systems theory as a science of feedback mechanisms, with the intention of understanding and ultimately controlling natural processes. This positivist and reductive approach to ecology remains strong to this day. In this mode, ecology is instrumentalized as a science and functions to monitor nature's overload capacity and adjust feedback mechanisms to enable continued development. Over time a tension developed between two approaches to ecology: one wherein nature is conceived as 'resources' to be managed and another where ecology as a science of complexity associated to a worldview oriented towards preserving the commons.

Ecology within the positivist paradigm oriented towards management of resources assumes its methods can help human control and predict ecological processes. While many of the concepts from this approach to ecology are useful for environmental assessment,[6] these approaches have limitations that are not always recognized including their inherent reductionism. Ecological theorist Stephan Harding explains that reductionist science remains useful 'if we want to design things like cars and computers, but its success is more limited in areas such as biology, ecology or in the realm of human social life where complex, non-linear interactions are the norm' (2006, 32). When linked to philosophical and political ideas, ecological thought is fraught with difficulties within the framework of positivist scientific rationality – as exemplified by the totalizing tendencies of some ecological theory. This problem arises because this type of scientific 'rationality' itself can be understood as a totalitarian model 'in as much as it denied rationality to all forms of knowledge that did not abide by its own epistemological principles and its own methodological rules' (Santos 2007, 16). As history demonstrates, references to nature can easily be used for oppressive political purposes. Nevertheless, ecological thinking and issues of justice can be complementary in theory and, in many instances, in practice (this is the case within many indigenous cultures and environmental justice movements). Merging the Western tradition of positivist science (which includes the conceptual framework that enables oppressions, i.e. patriarchy, racism, colonialism, etc.) with concepts of nature can enable hateful political philosophies. The reaction to these problems by the intellectual left has often been to ignore the ecological entirely, thereby avoiding the problem of naturalizing oppressive social constructs. In doing so, they created altogether new problems.

Ultra-Modernism

Postmodern theorists have emphasized the socially generated nature of knowledge, language and culture. Views of what constitutes natural behaviour or a normal environment are social beliefs. While being able to identify the constructed nature of particular social beliefs has helped to critically address how oppressions are reproduced in society, this focus on the social too often denies the material and the ecological. This critique of social constructivism is developed by ecological theorists who claims that it supports the 'existing parameter[s] of the modern worldview' (Spretnak 1997, 66). Since it fails to challenge the core discontinuities of modernity, post-modernist theory can best be understood as hyper-modernity or ultra-modernity rather than post-modernity (Griffin 1992; Spretnak 1997, 223; Sterling 2003, 222). A better critique of modernity (a truly post-modern alternative) 'would counter the modern ideological flight from body, nature and place' (Spretnak 1997, 223). Briefly and as a simplification, one might say that positivism denies the ecological by placing itself both outside of nature and in control of nature, while ultra-modernism denies the ecological by focusing exclusively on the social. Both can be characterized by six anti-ecological D's of modernity and ultra-modernity: disembodied, dis-embedded, de-contextualized, disengaged and disconnected and in denial of our own constitutive ecological self (Sterling 2003, 222).

Anti-environmental theory emerges from the cultural relativism of modernity. Science created a unifying worldview but simultaneously gutted cultural traditions. Nature was revalued and replaced with inert, mechanical and objectified matter. Its vitality was lost. Ultra-modernism accepts this logic. Its cultural relativism empties all points of view of moral significance. Bruno Latour claims that 'there was, however, a little hitch in the peaceful modernist version of politics: nature was as meaningless as it was disenchanted!' (2002, 11). This tradition has 'confused our relationship with the real' resulting in 'ideologies of denial' (Spretnak 1997, 5,8). The denial of the intrinsic value of non-human nature (e.g., the lack of moral significance to the extinction crisis) has enabled the rise of contrarian movements who seize upon ethical relativism to disengage with environmental concerns and launch anti-environmental campaigns. Disengagement leaves an ethical void where we need a platform for action.

Our culture is constructed and designers play a significant role in the construction of both material artefacts and immaterial ideas, concepts and sensibilities. But there is also an ecological context that underlies these constructions. The ecological has been so severely dismissed in our intellectual tradition that it will take a concerted effort to bring ecological ways of knowing back into our culture. The damage done by these denials is pervasive and now recognizably dangerous as many of our life sustaining ecological systems are severely degraded. The encounter with ecological realities in design must be an explicit priority. The arts have already been central to the reclaiming of the body and (to a lesser degree) ecological embeddedness. Design can work strategically towards this goal.

The Structure of Scientific Revolutions

Ecological theory posits that an reductionist and exploitative worldview is at the root of cascading ecological crises. This paradigm must be replaced by an ecologically coherent way of knowing. Thomas Kuhn's *The Structure of Scientific Revolutions* (1962) describes paradigm shifts in science as a change in the basic assumptions, epistemology and collectively held worldviews of a scientific community. Paradigms are the 'entire constellation of achievements – concepts, values, techniques, and so on shared by the members of a given community' (Ibid., 175). Since Kuhn's seminal work, the concept has been used to describe worldviews beyond the scientific community. Paradigms underlie ideologies and cultural assumptions.

The power of the concept of a paradigm is that it suggests that worldviews are maps, models and ways of knowing – rather than 'reality' itself. This idea was succinctly expressed in Alfred Korzybski's famous statement: 'the map is not the territory' (1933, 750). Once we acknowledge that paradigms are useful constructs rather than 'the territory' itself, we are in a position to judge the usefulness of a paradigm to our own ends. If a paradigm no longer serves our best interests, if inconsistencies need to be ignored and if the sum effects of our actions (based on the assumptions of the paradigm) are destructive, then it is time for a deeper analysis of our conceptual frameworks in order to find more appropriate models as a basis for action.

Paradigms are a way of perceiving associated with a collective worldview. Paradigm shifts are preceded by times where 'one can appropriately describe the field affected by it as in a state of growing crisis ... generated by the persistent failure of the puzzles of normal science to come out as they should' (Kuhn 1970, 67–68). Ecological theory describes the old Cartesian empirical/ reductionist/ mechanical paradigm as insufficient:

> The case against the dominant western worldview is that it no longer constitutes an adequate reflection of reality – particularly ecological reality. The map is wrong, and moreover, we commonly confuse the map (worldview) with the territory (reality). (Sterling 1993, 72 quoted in Sterling 2003, 119)

An ecological paradigm is potentially emergent. This paradigm does not negate positivism or social constructionism but it asserts the partial validity of reductionist science and the importance of social constructionist analysis – but it challenges these paradigms as the exclusive mode of knowledge.

> Western societies are experiencing the emergence of what can be termed a revisionary ... ecological paradigm, a fragile quality of 'third order change' or learning which offers a direction beyond destructive tendencies of modernism, and the relativistic tendencies of deconstructive post-modernism. (Sterling 2003, 115)

The awareness of paradigms as maps of reality rather than reality itself enables epistemological flexibility, meaning a capacity to discern which epistemic models and types of logic are most appropriate for a particular problem at hand. For example, clearly solutions to a mathematic problem are very different from the solution to an emotional problem. Different types of logic apply to different types of problems – so designers must be aware, flexible and prepared to apply different problem-solving approaches to different types of problems.

Post-normal Science

Post-normal science refers to a methodology of science that acknowledges its social context as well as the limitations of the scientific method. Since scientific work leads to technological and industrial innovations that have increasingly powerful and potentially destructive capacities, scientific decision-making must be incorporated into democratic decision-making processes. Post-normal science is science that understands itself functioning within a space where 'facts are uncertain, values in dispute, stakes high and decisions urgent' (Funtowicz & Ravetz 2003, 1). This relatively new modality recognizes that science must enter into a dialogue with other forms of knowledge and work towards practical, democratic and precautionary solutions in collaboration with others, as an extended peer community (Ravetz 2005). Post-normal science accepts the legitimacy of a plurality of perspectives and collaborates with other types of knowledge such as local, tacit and traditional ecological knowledge. This participatory approach counters the technocratic shortcomings of the scientific method and makes a definitive break with positivist science's tendency to present its own conclusions as value-free and universal. Post-normal science was developed as an outcome of the epistemological and methodological challenges in the sciences over the past century. In 1925, Alfred North Whitehead described a 'fallacy of misplaced concreteness' (1925, 64, 72) in science (the error of mistaking the abstract for the concrete). This fallacy is evident when scientific methods lead to an unwarranted illusion of certainty that is too often associated with the hard sciences. Today, post-normal science offers a methodology of science that is aware of its own limitations and its socio-ecological context, more democratic and mindful of the dangers of unintended consequences.

The Crisis of Reason

The specific ways that ecological context is dismissed and denied have been described in detail by philosopher and ecofeminist Val Plumwood in her seminal book *Environmental Culture: The ecological crisis of reason*. Plumwood describes the systemic devaluing of non-human nature in modernist thought where rationalist frameworks have led to the universal underestimation of complexity. The result is 'a cult of reason that elevates to extreme superiority a particular narrow form of reason and correspondingly devalues the contrasted and reduced sphere of nature and embodiment' (2002, 4). The dualistic split between nature/culture and body/mind has exalted culture and mind at the expense of nature and body. Whereas the rationalist explanation of the environmental crisis assumes that it is our 'nature' that has caused the crisis (greed, etc.) and it will be reason intensified (improved 'rational' markets) that saves us (Ibid., 6), Plumwood's critique holds that rationalist logic in this tradition is at the root of the environmental crisis. Only a more inclusive and holistic form of reason will help. From this perspective, markets are neither inherently rational or value free, but serve the interests of particular constituencies (with the design of systems of ownership and trading). The crisis of reason is propelled by habits of thought through which nature is devalued and opened to exploitation. Plumwood lists the ways in which the crisis of reason is propelled: through backgrounding, remoteness, instrumentalization and disengagement. I have added quantitative reasoning to this list. Ecological rationality challenges these modernist conceits and enables a more inclusive form of reason. With this analysis, designers can strategically address the various ways in which 'the crisis of reason' is perpetuated.

Backgrounding

Backgrounding is the process through which the activity and agency of nature is denied. It is the 'profound forgetting' of nature leading to 'dangerous perceptual and conceptual distortions and blindspots' (Plumwood 2002, 30, 100). Backgrounding happens through a failure to pay attention. To the disengaged it can appear as if resulting problems are not intentional. For example, it is not so much that ecological limits are misrepresented, 'but that limits are simply not recognized or investigated' (Ibid., 26). Backgrounding results in situations where 'nature's needs are systemically omitted from account and consideration in decision-making … we only pay attention to them after disasters occur' (Ibid., 108). As such, backgrounding is the mechanism that negates responsibility. Backgrounding enables the identification of 'the biospheric other as passive and without limits' (Ibid., 16). It gives modernity its rationalist hubris and overconfidence. By failing to see 'nature as collaborating partner or to understand relations or dependency on it' (Ibid., 30), humankind develops the illusion of autonomy, anthropocentrism and radical individualism – based on the false premise that any of us can survive independently of the ecological context. Backgrounding can be addressed by communication design that foregrounds environmental consequences of developments.

Remoteness

Remoteness refers to the distant consequences of our actions in highly complex industrial society where people remain oblivious to the consequences of their ways of living. Remoteness takes different forms, the most relevant being spatial, temporal and technological. Spatial remoteness is the physical distance between cause and effect. Temporal remoteness refers to consequences that are not felt for some time. Greenhouse gases emitted now take decades to change the climate and the warming effect will last centuries. Technological remoteness refers to the technological advantages that allow powerful groups to avoid the impact of environmental harms. Remoteness is a result of poor communicative and feedback links between actions and effects. Like backgrounding, remoteness feeds the illusion of disembeddedness and supports the denial of ecological realities. Remoteness protects decision-makers and the wealthy from direct environmental harms (because they typically live in places far removed as possible from ecological impacts) and even the awareness of ecological harms (as the communication systems do not work to alert everyone of all relevant environmental impacts). Remoteness is a problem that can be addressed by communication design strategies wherein causality, complexity and relations are made visible (see pp. 123–127).

Instrumentalization

Instrumentalization refers to the use of natural resources as utilitarian instruments for human consumption as opposed to the recognition of non-human nature as having intrinsic value outside its use to humankind. Instrumentalization distorts our sensitivity to nature and 'produces a narrow type of understanding and classification that reduces nature to raw materials' (Plumwood 2002, 109). Plumwood explains that 'in an anthropocentric culture, nature's agency and independence of ends are denied, subsumed in or remade to coincide with human interests, which are thought to be the source of all value in the world' (Ibid., 109). The objectification and 'the pacification of the objectified is a prelude to their instrumentalisation, since as a vacuum of agency, will and purpose, they are empty vessels to be filled with another's purpose and will' (Ibid., 46). The anthropocentric, ecocidal logic that assumes that non-human nature has no value outside of its use to humans enables strategies wherein 'others' are denied basic rights or even the right to exist at all. This othering strategy is a foundation for exploitation. Qualitative approaches to representing the natural world help counter its instrumentalization.

Disengagement

Disengagement is the ignoring of environment problems. It is the dominant public face of modernist rationality. Disengagement poses as neutral but is easily co-opted by economic forces. The disengaged present themselves as honest and unbiased but:

> Power is what rushes into the vacuum of disengagement; a fully 'impartial' knower can easily be one whose skills are for sale to the highest bidder, who will bend their administrative, research and pedagogical energies to wherever the power, prestige and funding is. Disengagement carries a politics, although it is a paradoxical politics in which an appearance of neutrality conceals a capitulation to power (Ibid., 43).

Feminist and other anti-oppressive theorists have described how the disengaged aim to appear rational and professional but what they reveal in fact is an absence of care. Plumwood describes the damage done by 'cloaking privileged perspectives as universal and impartial, and marking marginalized perspectives as "emotional", "biased" and "political" ' (Ibid., 44). Disengagement is perpetuated through 'well-practiced conceptual and emotional distancing mechanisms which legitimize the exploitation' (Ibid., 44). Today disengagement is perhaps the greatest tactic for denial of ecological reality. The façade of neutrality conceals complicity with the status quo and associated destructive industrial development. Designers can encourage engagement by being explicit about the politics of their practice and by highlighting the implications of depoliticized design.

Quantitative Reasoning

Traditional rationality is based on the dominance of quantitative reasoning that reflects empiricism's reliance on mathematical models (Santos 2007, 8). Positivist science converts phenomenon into numbers in order to study the subjects under investigation and develop systems of management. While quantitative reasoning has obvious virtues its hegemonic importance has dramatic consequences in the manner that the society is organized. Sociologist Boaventura de Sousa Santos explains:

> To know means to quantify. Scientific rigour is gauged by the rigour measurement. The intrinsic qualities of the object, so to speak, do not count, and are replaced by quantities that can be translated. Whatever is not quantifiable is scientifically irrelevant ... [thus] the scientific method is based on the reduction of complexity. (2007, 18)

The reliance on quantitative reasoning inhibits the perception of depth and intrinsic value by diminishing the complexity of nature and flattening out phenomena to what can be captured by numbers. This sense of being able to quantify all objects of study, the single-minded faith in quantitative methods and ignoring of qualitative values reduces life to abstractions. Properties of complex adaptive systems such as emergence, resilience, adaptability and robustness are properties that cannot be quantified reliably and yet are fundamental to long-term health and sustainability. For this reason, quantitative methods will always have limitations. While clearly quantitative methods and logic are valuable approaches for many problems, it is the dominance and over-reliance on these approaches that is the problem as it often replaces more nuanced understanding of the full set of relationships within eco-social phenomena.

Ecological Rationality

Each of us exists within the environment that enables our lives. And yet normative conceptions of nature present this environment as ripe for abuse. The modernist 'machine of reason depends on what it destroys for its survival. Its rationality is ultimately suicidal' (Plumwood 2002, 236). While 'no rational society rewards members to undermine its existence' (Orr 1992, 6), in the current context ecologically exploitative behaviour is often encouraged. All too often it is profitable. Ecological theory suggests that the 'contrived blindness to ecological relationships is the fundamental condition underlying our destructive and insensitive technologies and behaviour' (Plumwood 2002, 8). As ecological beings the consequences of the denial of the ecological are personal (in the form of psychological illness and pathologies); social (in the form of ecologically destructive behaviour) and environmental (climate change, etc.). Ecological theorists such as Plumwood and Orr propose a better form of reason where attitudes and actions are matched with survival claims. As a remedy to the intellectually incoherent crisis of reason described in this chapter, ecological theory offers a more thorough form of rational thought. Ecological epistemology, ontology, ethics, literacy, design, movements, perception and identity complement ecological rationality. These will be explored throughout the rest of Part 2.

Notes

1. Some content in this chapter (and other chapters) come from ideas developed in my PhD (2012) and in other published work (Boehnert 2017). This book expands on ideas presented earlier with an elaboration of an ecological theory for design.
2. Epistemology is the study of the nature of knowledge and ways of knowing.
3. Ontology is the study of the nature of being.
4. Smut's legacy on ecological thought was featured in Adam Curtis' *Machines of Loving Grace* (BBC Two 2011). This documentary exposes how 'holism became a tool to make the British Empire more stable' and how 'theories of nature are highly politically charged' (quote by Peder Anker, a historian of ecology, interviewed in the film). Curtis explains 'what Smuts was doing showed how easily scientific ideas about nature, and natural equilibrium, can be used by those in power to maintain the status quo'. The documentary presents instrumental approaches to ecological thinking (pioneered by Jay Forrester, Howard Odum and Buckminster Fuller) and fails to capture the self-reflective insights from Gregory Bateson and others in the 'soft' systems approach. This partial coverage in a populist style documentary has the effect of leaving audiences with a misunderstanding of the varieties and subtleties of ecological theories and associated debates.
5. This problem leads to charges of 'eco-fascism': a loaded term reflective of assumptions of what constitutes 'freedom' and who gets to enjoy freedom when one's activities harm other people and the environment. Which people, other living beings and what natural processes have the 'right' to freedom if this denies freedoms (and even rights to exist) to others? Who and which parts of nature will be protected against destructive human activities and industrial processes?
6. Ecological assessment tools, strategies and concepts include concepts such as carrying capacity, ecological footprint, ecosystem services and planetary boundaries.

5 Epistemology Error

When things are not working properly, it is often necessary to look into the philosophical roots of our habitual practices. Our understanding of reality, way of knowing and epistemology leads to particular types of practice in business, finance, culture, education, politics and design. When our ideas are not aligned with the ways that the world actually works, dysfunction occurs. The concept of epistemological error was developed by Gregory Bateson in his seminal book *Steps to an Ecology of Mind* (1972). Bateson explains that our map of reality is a poor reflection of reality itself; 'most of us are governed by epistemologies we know to be wrong' (1972, 493). Over the course of his career, Bateson laid the foundations for contemporary ecological thought with insights in ecological philosophy, communication theory, aesthetics, mental health and perception. His prolific work at the intersection of psychology, cybernetics, philosophy, anthropology and natural science focused on communication, philosophies of mind and human–nature relations.

While the notion that the dominant contemporary worldview is a poor reflection of reality has been described by cultural commentators in multiple fields (Bertalanfry 1969; Capra 1982, 1997, 2003; Shiva 1988, 2005; Bohm 1992; Orr 1992; Plumwood 2002; Santos 2007; Meadows 2008), Bateson was the first to use the term 'epistemological error' and offers some of the most articulate descriptions of the nature and gravity of this condition:

> I suggest that the last 100 years or so have demonstrated empirically that if an organism or aggregate of organisms sets to work with a focus on its own survival and thinks that is the way to select its adaptive moves, its 'progress' end up with a destroyed environment. If an organism ends up destroying its environment, it has in fact destroyed itself. (Bateson 1972, 457)

The error in thought is deep reaching. The theory of epistemological error posits that the Western premise of radical independence from non-human nature is wrong. As ecological beings, we are embedded and mutually dependent on the rest of the natural world but our understanding does not reflect these basic geophysical and biological circumstances. Consequently, we have constructed deeply unsustainable ways of living.

Epistemological error refers to a way of knowing that is the legacy of the Western philosophical tradition. It determines that we generally do not perceive systemic interconnections and are thereby unable to deal with the complexity presented by converging ecological, social and economic crises. It is not that we innately cannot deal with interconnectedness and interdependence, but that this reality is effectively hidden by the complexity of contemporary conditions and our inadequate epistemological tradition. Humankind has conceived of itself as the sole proprietors of sentience and the rest of the world 'as mindless and therefore as not entitled to moral or ethical consideration' (Ibid.,

62). The narrowing down of our epistemology to reflect only our own interests (or even the interests of our own species) and the instrumental processes we use to do this are at the root of current problems.

> When you separate mind from the structure in which it is immanent, such as human relationship, the human society, or the ecosystem, you thereby embark, I believe, on fundamental error, which in the end will surely hurt you. (Ibid., 493)

Humankind is collectively ignoring information vital for our survival. Premises are disconnected from actual circumstances: rational logic is in conflict with its context. Human ways of knowing evolved towards instrumental ends to serve human desires, without taking into account ecological embeddedness. The reductive focus is ultimately self-defeating and irrational.

> When you narrow down your epistemology and act on the premise 'what interests me is me or my organization or my species', you chop off consideration of other loops of the loop structure. You decide that you want to get rid of the byproducts of human life and that Lake Erie will be a good place to put them. You forget that the ecomental system called Lake Erie is a part of your wider ecomental system – and that if Lake Erie is driven insane, its insanity is incorporated in the larger system of your thought and experience. (Ibid., 460)

Bateson describes the narrowing down of our epistemology and the failure to perceive ecological embeddedness as the result of this system of erroneous thought: 'there is an ecology of bad ideas, just as there is an ecology of weeds' (Ibid., 492). These 'bad ideas' are particular habits of mind and ways of perceiving things in isolation. These habits emerge from an intellectual tradition and will continue to be reproduced unless we make a concerted effort to change tack.

These errors are manifest in individual and social behaviour. Humankind's 'massive aggregation to man and his ecological system arise out of efforts in our habits of thought at deep and partly unconscious levels' (Ibid., 495). While epistemological error is not necessarily a serious problem 'up until the point at which you create around yourself a universe in which that error becomes immanent in monstrous changes of the universe that you have now created and try to live in' (Ibid., 493), in a technologically advanced society epistemological error is lethal. The basic epistemological fallacies are:

1 that mind and nature are independent of each other;
2 that humans are separate from the rest of the natural world;
3 that this separation creates a 'natural' and necessary competition;
4 that humans must strive to dominate each other and non-human nature;
5 that instrumental approaches to knowledge can describe an objective truth; and
6 that the non-human natural world has no agency of its own.

The concept of epistemological error suggests that humankind is undergoing a crisis of perception, based on misperception. This misperception involves a basic failure to perceive mutualistic ecological relations. These fragmenting epistemological premises are a dysfunctional legacy of modernity. Sustainability education theorist Stephen Sterling explains that 'the dominant western epistemology, or knowledge system, is no longer adequate to cope with the world that it itself has partly created' (2003, 3). In a response to this dilemma, Sterling advocates a shift 'from the machine metaphor to the systemic metaphor of ecology' (Ibid., 8). Ecologically coherent epistemologies are a necessary basis for lucid rationality and ethics.

The things we make are extensions of our ways of knowing. Epistemological error is encoded in the language we use, the objects we create and the cities we build. Bateson warns of the self-validating power of ideas: the world 'partly becomes – comes to be – how it is imagined' (1980, 223). Epistemological error is designed into cultural artefacts, language and systems making it exceedingly difficult to fix. It is reproduced in education, communication, media, policy, law and design. System structures, social institutions, architecture, designed artefacts, media and habits of mind all reinforce it. Many of us feel stuck because our way of knowing determines that we are incapable of perceiving interrelationships and thus ill-prepared to even conceive of solutions.

Bateson's own work on communication levels offers a foundational framework to address this dilemma. Groups who have faced oppression have developed strategies that offer useful models for revealing epistemological blind spots. Most of us are already moving beyond epistemological error.

Alternative Epistemologies

Feminist, class, race, indigenous and decolonization scholars and activists have all contributed to the critique of modernity and the ways in which its assumptions privilege the interests of certain groups at the expense of others. Anti-oppressive work has demonstrated how conceptual frameworks reflecting the interests of powerful people are used to justify exploitation of both people and the non-human natural world. For this reason, feminist and other anti-oppressive theorists' contributions are foundational to the epistemological work that needs to be done for the environment.

Feminist Challenges and Contributions

Intersectional feminist theory contributes a profound critique of androcentrism within modernity and its supposed value neutrality. Over the past few decades, women of colour have created a more inclusive type of feminism that interrogates the intersections of oppression. The inclusion of women's experiences and knowledge within our epistemological tradition contributes to richer, more relational, situated, experiential, empathetic, caring and generally diverse ways of knowing. Feminist analysis reveals the unjust conceptual frameworks, social conditions and institutional practices supporting domination: 'Injustice does not take place in a conceptual vacuum, but is closely linked to desensitising and othering frameworks' (Plumwood 1999, 197). 'Othering' refers to dismissing and denying a group the same considerations and rights as those with greater privilege. Othering has been a means to justify the domination of women, people of colour and other groups. Building on feminist thought, ecofeminists extend the critique of the subjugation of women to that of the natural world by describing how similar intellectual strategies are for both types of domination.

Feminist analysis places the historical roots of domination as emerging at different times: Raine Eisler's *The Chalice and the Blade* (1988) as the sixth to third century B.C.; Carolyn Merchant's *The Death of Nature* (1980) as between the sixteenth and eighteenth century; and Val Plumwood's *Environmental Culture: The Ecological Crisis of Reason* (2002) with the tradition of rationality in the West – starting in classical Greece but coming to a fruition in the scientific revolution (fourth century B.C. to eighteenth century). Each era contributed to the logic of domination that was apparent in the basic beliefs, values, attitudes and assumptions towards women that the feminist movement resisted, critiqued and partially transformed over the past 150 years. Ecofeminists claim that the ways in which women's experiences, subjectivities and interests have been historically denied parallel the ways in which the interests of non-human nature are denied. Social injustices emerge from desensitizing and othering frameworks in similar ways to environmental injustices. If this is true, then feminist strategies that have worked to help women transform power relations and make women's needs and interests visible are of use with the wider environmental movement.

Feminist and ecofeminist scholars describe experiential, situated knowledge and emotional intelligence as central to resolving the false dichotomy between intellect and body, humanity and nature. The feminist emphasis on the experience of a subject in a particular situation places value on the knowledge that is embedded in customs, practices, languages and cultures. 'Situated knowledge,' a term coined by Donna Haraway, is about the location, position and site of rational knowledge claims (1988). Claims are best understood as a view from an individual in particular situation. Knowledge is best understood when it is situated and also when it take emotions into account. Karen Warren cites Daniel Goleman's work on emotional intelligence as significant in its recognition that emotions matter for rationality; 'the intellect (rational mind) simply cannot work effectively without emotional intelligence' (Goleman 1995, 28 quoted in Warren 1999, 135).[1] Warren argues that without emotional intelligence we have no capacity for moral reasoning (1999, 136). Values advocated by feminists (such as care and empathy) have been central to social change over the past century and are also integral to ecological thought. But feminism has far more to offer than simple advocacy of kinder values. Black feminist Angela Davis calls for an abolitionist feminism that seeks to abolish all forms of sexist and racist violence along with the social structures that reproduce these violences – including capitalism. The aim of this radical feminism is not to take power from men but to transform the structure of leadership and society as a whole for the benefit of women and men and all races and sexual orientations.

EPISTEMIC SELECTIVITIES, PRIVILEGE AND BLINDNESS

An ecological worldview is a major break with the dominant epistemological tradition and is still fiercely contested in places where power is situated. In response to this problem, ecological theorists have described epistemic selectivities as 'mechanisms inscribed in political institutions which privilege particular forms of knowledge, problem perceptions, and narrative over others' (Brand & Vadrot 2013, 218). Epistemic selectivities reproduce instrumental logic and unsustainable development and obscure options outside this perspective. Since feminists face similar challenges with our own marginalization and the dismal of our perspectives, the strategies feminists use to overcome our own epistemological ostracism can help confront and disrupt hegemonies that deny the interests of non-human nature.

The concept of epistemic privilege describes how the experience of oppression gives individuals especially clear insights on how power functions in society. Being subject to domination requires individuals to pay attention to the perspectives and expectations of those with power (to be aware of potential punishments or rewards). Being on the receiving end of oppression makes one especially attuned to the power imbalances and power dynamics of a situation. The privileged do not need to consider the perspectives of those with less power or the power dynamics around these relationships and can concern themselves with more entertaining activities without fear of reprisals. Having the lived experience of being badly treated can also build an acute sense of injustice and a desire to fight injustices that are more clearly perceived from this position. On the other hand, the powerful often remains unaware of the ways in which their activities and attitudes oppress others (and they have an interest in remaining unaware and avoiding any associated guilt). The class, race and gender with the most privilege are least likely to notice how oppression works, because they benefit from current power dynamics and they don't need to pay attention to these dynamics to survive. Epistemological blindness enables some to remain blissfully ignorant of the destructive impact of vastly unequal power relations while it also limits their understanding of the politics and power dynamics of complex problems – such as climate change.

Epistemological privilege is associated with feminist standpoint theory (see p. 170) and Sandra Harding's concept of strong objectivity that suggests that biases can never be removed from truth claims, and that marginalized perspectives can create a stronger objective point of view (1986). While the danger of romanticizing the least powerful is a real one (and this concept should not be understood to imply that a particular argument by a marginalized person is necessarily correct), these ideas help build a case for more inclusive approaches to knowledge that approach power critically. Obviously these are highly contested ideas, especially with people who come from elite backgrounds and institutions.

Indigenous Epistemologies and Epistemologies of Colour

The ongoing legacy of colonization and racism continues to take its toll on the lives of individual people and society at large as it enables distinct injustices and also marginalizes the interests and outlooks of less advantaged groups. Where the interests and subjectivities of the privileged class, race, sexual orientation and sex are presented as universal, the contribution of other perspectives is denied. Before focusing on contributions of brown, black and indigenous people's cultures to environmental debates, I review some of the challenges they face advancing their ideas and interests.

COLONIZING AND DECOLONIZING EPISTEMOLOGIES

Western colonialism relies on the dismissal of truth claims and ideas that disturb the epistemological hegemony of the dominant cultural tradition. Women such as Indian ecologist and anti-GM activist Dr Vandana Shiva regularly deals with such institutionalized prejudice. This was epitomized in Michael Specter's feature-length character assassination in *The New Yorker* (2014) dismissing her anti-GM campaigning. Shiva's argument against GM agriculture in India presents claims that challenge Western hegemony – and the corporations that secure and maintain current power structures. The debate that followed illustrates the validity of Bruno Latour's belief that the West will accept difference, with a 'combination of respect and complete indifference' (2002, 15) as long as other cultures do not claim 'any ontological pretensions' (Ibid., 16). Thus,

> Diversity could be handled by tolerance – but of a very condescending sort since the many cultures were debarred from any ontological claim to participate in the controversial definition of the one world of nature. Although there could be many warring parties engaged in local conflicts, one thing was sure: there was only one arbiter, Nature, as known by Reason. (Ibid., 9)

Here views from the Global South that contest the practices and power of Western corporations are ridiculed under the guise of rational discourse. Yet only one side of the debate was able to engage with *The New Yorker*'s audience (as the magazine did not publish the statement Shiva wrote in her own defence or refer to any of the plentiful research that would have supported Shiva's position). This type of one-sided journalism is a none too subtle means of dismissing perspectives that do not align with Western perspectives and, in this case, imperialist interests.

Anti-oppression scholars and activists have different means for pushing forward their agenda. Their invaluable contribution is the articulation of the value in other ways of relating to each other and the planet. In 1984, black feminist Andre Lorde stated:

> It is learning how to take our differences and make them strengths. For the master's tools will never dismantle the master's house. They may allow us temporarily to beat him at his own game, but they will never enable us to bring about genuine change. (2007 [1984], 111)

Black women's experiences of being relegated to 'western modernity's nonhuman other' (Weheliye 2014 quoted by Frazier 2016, 41) and with an 'illusory subject/object status [that] has always already paved the way for their extreme instrumentalization' (Frazier 2016, 69) inform a deep-reaching critique of Western colonial assumptions. Focusing on black female subjectivity, Chelsea Frazier describes black feminist interventions in environmentalism as 'troubling ecology' in ways that disrupts environmental studies frameworks and offer alternative conceptions of ecological ethics (2016, 40). Frazier argues

> that 'the West' itself – its divisions of space and its rigid notions of the human subject – are insufficient frameworks through which 'global warming, severe climate change, and the sharply unequal distribution of the earth's resources' can be effectively addressed. We must consider these issues while concurrently addressing a central conflict from which these issues emerge: a fraught and delimited understanding of human subjectivity. (2016, 44)

Drawing on Jane Bennett's notions of vital materialism, with its insistence of the vitality or aliveness of all matter, its 'sense of a strange and incomplete commonality with the out-side' (Bennett 2010, 61 quoted by Frazier 2016, 68) and its understanding that 'all bodies become more than mere object' (Bennett 2010, 13 quoted by Frazier 2016, 45), Frazier argues for a vital materialist stance as superior to an environmental one. But she advocates for a 'troubling of that vitalist material ethic' (Ibid., 69) informed by the black experience of 'discursive objectification' (Ibid., 69). This troubling is contingent on de-stabilizing and reshaping the hierarchies, classifications and 'visual, spatial and philosophical assumptions' (Ibid., 69). It foreshadows an environmental politics based on a deeply restructured human subjectivity, beyond 'hierarchical myopia and politics of exclusion that have plagued environmental discourses' (Ibid., 68). These insights stretch the ecological imagination as they sketch new modes of relating the space we inhabit.

TRADITIONAL ECOLOGICAL KNOWLEDGE

Marie Battiste, a Canadian First Nations scholar claimed: 'For as long as Europeans have sought to colonize Indigenous peoples, Indigenous knowledge has been understood as being in binary opposite to "scientific", "western", "Eurocentric" or "modern" knowledge' (2002, 5). Traditional Ecological Knowledge (TEK) is shared by diverse indigenous communities and has supported sustainable existence for these people for at 7,000 indigenous nations around the world over 40,000 years (Hughes 2012, 14). The epistemologies that under girth these societies are valuable sources of learning on how to conceive of human well-being and functional relationships with the non-human natural world. Despite or perhaps because of this explicit connection to the natural world, indigenous people often encounter prejudice when describing their own experiences and culture. Indigenous cultures have cultivated place-based knowledge that 'are an expression of generations of co-evolution of humans and the ecosystem they inhabited' (Wahl 2016, 159). Over the years, these cultures have often increased bioproductivities of their bioregion (Ibid., 158; Erle et al. 2012) in ways that can now inspire regenerative design. Native American scholar and activist Vine Deloria claims that within 'the tradition, beliefs and customs of the American Indian people are guidelines for preserving life and the future of all nature' (quoted in Lauderdale 2007, 739). Indigeneity offers a rich archive of ecological epistemological traditions, including jurisprudence based on 'substantive reliance of interrelatedness of nature' (Ibid., 741). Indigenous scholars and ecological theorists often agree that TEK provides inclusive and equitable approach the numerous crises (Lauderdale 2008, 1836). Western science is slowly recognizing the value of ethnoecology for example.

Meanwhile, indigenous people are rightfully wary of having their knowledge stolen. Western drug companies have patented plants that have been cultivated by indigenous peoples over thousands of years. Vandana Shiva is one of many fighting biopiracy (theft of property, tradition knowledge and biological/genetic resources) as a modern form of colonization. Indigenous peoples have traditionally fiercely rejected the idea that land could be privately owned since they typically see all natural resources as belonging to the community (Hughes 2012, 51). For this reason patenting indigenous plants for private profit for outsiders, especially in the context of the violence of colonialism, is a deeply controversial and offensive practice to many indigenous people.

INDIGENOUS EPISTEMOLOGIES OF VITALITY AND KINSHIP

The common ground for indigenous people around the world is evident in their traditions commonly characterized by a kinship and respect for non-human nature and its vitality. Indigenous world views see 'the world as alive' and our relation to 'the rest of life is one of participation' (Wahl 2016, 159). Human ecologist Alastair McIntosh describes why Western academics and journalists find indigeneity so hard to accept:

essentialism is anathema equally to reductionist forms of modernity and deconstructionist postmodernity ... To some secular rationalist thinkers spiritual essentialism is the royal road to Nazism (Biehl 1991, 100–101), the logic being that because the Nazis used essentialist notions of German identity this means that all essentialism teeters on the edge of totalitarianism. Such thinking is as sloppy as it would be to blame surgeons for knife crime. The challenge that premodernism poses to post/modernity is therefore grave. It considers some of the most paradigmatic thrusts of post/modern thought – those which, in their arid materialism, deny the spiritual *esse* – to be violations of Being. (2012, 42–43)

In turn, indigenous activists critique European culture as exploitative and violent in its intellectual tradition. In a controversial speech given at the Black Hills International Survival Gathering in 1980 Native American Russell Means declared that Western industrial society 'seeks to "rationalize" all people in relation to industry maximum industry, maximum production. It is a doctrine that despises the American Indian spiritual tradition, our cultures, our lifeways' (1980) (Means explicitly includes the radical Marxist left-winged tradition in this critique). He maintains:

> The European materialist tradition of despiritualizing the universe is very similar to the mental process which goes into dehumanizing another person ... the dehumanizing is what it makes it all right to kill and otherwise destroy other people ... In terms of the despiritualization of the universe, the mental process works so that it becomes virtuous to destroy the planet. Terms like progress and development are used as cover words here, the way victory and freedom are used to justify butchery in the dehumanization process. (1980)

The Native American tradition asserts that the Earth will retaliate abuse: 'American Indians have been trying to explain this to Europeans for centuries. But ... Europeans have proven themselves unable to hear' (1980). Unfortunately, as I describe in Chapter 6, learning theory indicates that even when people do hear, hearing does not always accompany learning – or the capacities to act on new knowledge (see pp. 78–80). Today researchers are finding pathways to sustainability in indigenous wisdom (Hendry 2014).

Indigenous cultures have created forms of jurisprudence wherein 'law and nature were bound together ... based on concrete notions of the individual and collective good than on a modern abstraction imposed by the nation-state as the ideal to which people must conform' (Lauderdale 2007, 739). Indigenous law embodies ideas of social diversity and emphasizes social responsibility as the cornerstone of law. Herein 'social organization and cultural values are based on learning the lessons of nature' (Ibid., 741). Indigeneity is rooted in core values based on communal life: 'indigenous people see everything through the filter of community' (Harris & Wasilewski 2004a, 494) and have ways of living based on inclusive and relational practices and creating dialogic space where 'there is no coercion in government, only the compelling force of conscience' (Ibid., 497). The core values shared by indigenous peoples can be described as 'four Rs of relationship, responsibility, reciprocity, and redistribution' (Harris & Wasilewski 2004a, 492–492). Reciprocity is 'based on very long relational dynamics in which we are all seen as "kin" to each other' (2004a, 493). The literature suggests that kin is a central concept in indigenous epistemologies.

Robin Wall Kimmerer, director of the Center for Native Peoples, explains that 'in Anishinaabe and many other indigenous languages, it's impossible to speak of Sugar Maple as "it". We use the same words to address all living beings as we do our family' (2015).

> In indigenous ways of knowing, other species are recognized not only as persons, but also as teachers who can inspire how we might live. We can learn a new solar economy from plants, medicines from mycelia, and architecture from the ants … Among the many examples of linguistic imperialism, perhaps none is more pernicious than the replacement of the language of nature as subject with the language of nature as object.
> (Ibid.)

Kimmener proposes two new pronouns: 'ki' (singular) and 'kin' (plural) because 'words have power to shape our thoughts and our actions. On behalf of the living world, let us learn the grammar of animacy. We can keep "it" to speak of bulldozers and paperclips, but every time we say "ki", let our words reaffirm our respect and kinship with the more-than-human world' (Ibid., x). Language objectification. New words make space for new types of relationships. Other indigenous languages offer similar insights. The Algonquin concept of *wetiko* describes 'a psycho-spiritual disease of the soul which deludes its host into believing that cannibalizing the life-force of others is logical and moral' (Bollier 2016). Like many indigenous concepts, *wetiko* add a dimension of spirit and soul to descriptions of human existence.

Academics and journalists are often quick to reject indigenous perspectives as essentialist, romanticist or simply new age tripe. But the new age movement is a Western phenomenon that appropriates knowledge from other cultures and makes superficial substitutions sometimes accompanied by dubious claims. This should not reflect badly on indigenous cultures but on the Western pillaging. With this brief summary of TEK epistemologies, I am not suggesting that all indigenous people have always been sound ecological stewards. Nor am I suggesting we all live in the various ways indigenous people have in the past or present. I am saying that there are other ways of knowing and local practices that have value and that this knowledge will only be accessible by acknowledging and respecting difference – rather than attempting to assimilate everyone into the European, American or other hegemonic traditions. One place where this work is already being done is in ecopsychology. In their quest to find better ways of understanding the psychological and the ecological, ecopsychologists have drawn inspiration from indigenous ways of describing human–nature relations (see pp. 138–142).

Ecological Epistemologies and Ontologies

Ecological theorists describe a dramatic change in epistemology and ontology away from modernist's conceptions of human–nature relations. Ecological epistemologies are under girthed by the awareness of ecological embeddedness and the subsequent revision of knowledge systems to reflect this awareness. Ecological ontology refers to our actual constitutive embeddedness within and as part of larger ecological systems. Nested systems are embedded inside each other (Feibleman 1954). Ecological systems come in many sizes each nesting within the next (from microscopic to planetary). Nested systems also describe the relationship between ecological, social and economic systems: the economic system is dependent on the social system that is dependent on the ecological system (see Figure 3.1, p. 42). The Earth system will carry on with or without humans (albeit in some degraded form i.e. significantly less biodiversity, dead seas, desertification, toxicity, etc.). Dysfunction occurs when the parts (i.e. human society/the social system) do not recognize themselves as nested within larger systems (i.e. ecological system). This illusionary autonomy results in pathological behaviour. This is the case with the current economic system, which does not acknowledge embeddedness or respect ecological limits (Daly 1972, 1996; Georgescu-Roegen 1972; Meadows et al. 1972; Brown 2008; Jackson 2009). This pathological behaviour is similar some respects to how a biological cancer creates disorder in the human body. The financial and economic systems create disorder at the social and ecological levels of life organization (McMurtry 1999). Ecological economic theory suggests these dysfunctional dynamics are due to an economic system that was not designed to take account of its own embeddedness within the social order and the ecological orders (see Chapter 3). Dysfunction in nested systems arises when nesting order is not acknowledged in the systems we design: often with dramatic social and environmental consequences.

Ecological Ethics

Early ecological theorist Aldo Leopold advocated an extension of ethics. All ethics, according to Leopold, are based on 'a single premise: that we are members of a community of interdependent parts' (2001 [1949]: 98). He proposed a simple ethic: 'A thing is right when it tends to preserve the integrity, stability, and beauty of the biotic community. It is wrong when it tends otherwise' (Ibid., 110).

> The land ethic simply enlarges the boundaries of the community to include soils, waters, plants, and animals, or collectively: the land ... In short, a land ethic changes the role of Homo Sapiens from conqueror of the land-community to plain member and citizen of it. It implies respect for his fellow-members and also respect for the community as such.
> (Ibid., 204)

The simplicity of this ethos is attractive. But in a complex world, ecological ethics are far from simple. The wider the boundaries of concern, the more difficult ethics become. Extended ethical boundaries of concern are a radical starting point. How this gets put into practice is where everything becomes exceeding complex, messy and political.

Ecological ethics are complicated by the remoteness of industrial processes, poor communicative links and unintended consequences. Unintended consequences are a characteristic of modernity (Giddens 1990) resulting in a seeming loss of ethical ability to act responsibly (since consequences are distant in time and space – or entirely unknown). The ethical response to unintended consequences is to attempt to understand their nature and exercise precaution (rather than to deny their existence altogether). Where technology develops faster than the ethical frameworks and social institutions to ensure unintended consequences are investigated and avoided, ethics are severely compromised.

The feminist contribution to the debate about ethics is evident through their 'articulation of values lost in mainstream ethics, e.g., the values of care, love, friendship, and appropriate trust' (Warren 2001, 334). Plumwood (2002) describes why respecting non-human nature's regenerative capacities and nurturing an ethic of care is essential not only for ethical reasons but as a basis for a coherent rationality. This ethic of care is consonant with the indigenous contribution based in relationships and kinship. It is no coincidence that all traditions that value the nurturing ethos are marginalized in this economic system that extracts value for capital accumulation and rewards exploitative behaviour.

Confronting Error

While epistemological error is deeply entrenched in contemporary thought, it is a way of knowing that has been learned – so it is also a way of knowing that can be challenged and reconstructed. Particular habits of thought have enabled the exploitation of the environment. Since the strategies used to 'other' nature are similar to those that have backgrounded the rights and needs of women, people of colour and indigenous cultures, ecofeminists and anti-colonization activists and scholars argue that anti-oppressive work is a useful starting point for understanding the conceptual mechanisms used to exploit non-human nature. The women's movement's partial success in re-orienting society around new values provides valuable insights for the environmental movement. Intersectional feminism has intellectual strategies as well as practical methods that have proven to work to help challenge, disrupt and transform hegemony. As individuals learn to identify, intervene and transform the ideas, practices and institutions through which the social order is reproduced, social change is evident. Anti-oppressive practices include an attention to ideology embedded in myth, images, narratives, cultural media and design.

These intersectional and decolonalization critiques of the various 'isms' have recently been introduced into design. With a boycott of the bi-annual Design Research Society conference in 2016, the newly established Decolonizing Design group published a statement:

> the field of design studies as a whole is not geared towards delivering the kinds of knowledge and understanding that are adequate to addressing the systemic problems that arise from the coloniality of power. We acknowledge that this deficiency is a reflection of the limitations of the institutions within which design is studied and practiced, as well as the broader context of the colonial matrix. (Ansari et al. 2016)

The group challenges assumptions embedded into the Anglocentric/Eurocentric tradition in design in ways that are often explicitly pro-environmental: 'designers can put to task their skills, techniques, and mentalities to designing decolonial futures aimed at advancing ecological, social, and technological conditions where multiple worlds and knowledges, involving humans and non-humans, can flourish in mutually enhancing ways' (Ibid.). Decolonizing scholars and activists excel at exposing the limitations of the current worldview and social order while building networks of solidarity to support the challenging work they are undertaking. Their deep-reaching critique of power and unsustainability highlights structural injustices and harnesses transformative practices towards deep reaching social change. Intersectional feminism is a building block for the next social transformation – that of enlarging the community of concern to include the wider ecological community.

Note

1. Goleman's later work *Ecological Intelligence* (2009) was significantly less impressive. I am mentioning it here as the idea of ecological intelligence could be similar to ecological literacy, but it is developed instead in a limited manner in terms of consumer choices (rather than a way of thinking that embraces and integrates ecological thought. In this formulation, ecological literacy and ecological intelligence profoundly challenge modernist assumptions).

6 Ecological Literacy

It should now be obvious that ecology is a science that has profound philosophical implications. In a 1964 paper titled 'Ecology – A Subversive Science', Paul Sears proposed that if ecology was taken seriously it would 'endanger the assumptions and practices accepted by modern societies' (11). Ecological thought breaks the epistemic error of modernity. It addresses fragmented consciousness and challenges intellectual constructs that justify the exploitation of nature. Design practice is still catching up with the insights of ecology theory. Part of the reason why the task of ecological learning is so severe is that it is not simply a collection of facts to be added onto what we already know but it is a kind of learning that requires an interrogation of many basic premises. For example, in light of the recognition of humankind's interdependence with our environment, what right does anyone have to make pollution that will destroy the well-being of others – now and in the future? Partially due to the acutely difficult nature of this type of question, ecological literacy remains marginal in design education and in design practice (as well as other disciplines and sectors). Since ecological learning fundamentally disrupts and challenges educational cultures and assumptions about what constitutes good design – there is intense institutional resistance. Nevertheless, it is no exaggeration to say that the future rests on our capacity to become ecologically literate and capable of designing ecologically sustainable ways of living.

The ambitious aim of ecological literacy is to create the frame of mind that recognizes the ecological and organizes cultural, political, legal and economic priorities accordingly. In a society with ever-increasing technological capacity for both beneficial and destructive industrial development, ecological literacy is what will make the difference. The tendency towards fragmentation in the Western intellectual tradition makes sustainability an impossible achievement when approached through reductive modes of analysis and the ensuing focus on highly individualistic consumer choices and nudging behaviours. Ecological literacy addresses the problems that emerge from a way of knowing characterized by radical discontinuity with non-human nature. Acknowledgement of geophysical relationships is a foundational step towards transforming learning and cultural priorities. Ecological literacy envelops the tradition of modernity into a more inclusive and functional worldview. In this chapter, I present a theoretical introduction to ecological literacy. I review a theory of levels of learning that informs moving from theory to practice and addressing the value/action gap in sustainably education and communication. I propose two modes of ecological literacy as a response to its continued marginality in institutional contexts and the need for critical and systemic approaches.

David Orr coined the concept of ecological literacy in his influential work *Ecological Literacy* (1992) where he describes why ecological understanding must become a pedagogic priority across all disciplinary traditions. Education must impart an understanding of the interdependence between natural processes and human ways of living. Ecological literacy demands a type of education that nurtures the capacity to think broadly. We must 'reckon with the roots of our ailments, not just their symptoms' (Orr 1992, 88). The depth and scope of the ecological problem is linked to how we think:

> The disordering of ecological systems and of the great biogeochemical cycles of the earth reflects a prior disorder in the thought, perception, imagination, intellectual priorities, and loyalties inherent in the industrial mind. Ultimately, then, the ecological crisis concerns how we think and the institutions that purport to shape and refine the capacity to think. (Orr 2004, 2)

In a technologically advanced society, understanding the ecological impacts of our ways of living is necessary for informed citizenship and basic ethical coherence. But ecological literacy is not only about understanding the ecological impact of actions, but about developing new capacities to act and create ecologically viable ways of living over time.

Ecological Literacy versus Sustainability

Ecological literacy challenges the abused and ambiguous term 'sustainability'. Sustainability as a concept is often rendered relatively meaningless through its use by businesses keen to appear to be doing the right thing in meeting corporate social responsibility (CSR) objectives and not implicated in the escalating ecological crisis. Although various aspects of sustainable practice can be measured using environmental assessment processes, the lack of rigorous standards combined with the failure to adjust boundaries of concern wide enough to include the full impact of industrial processes, results in rampant misuses of the term. Frameworks for making ecological assessment legally binding or holding CEOs and industrialist leaders morally and legally accountable are typically weak or even non-existent. Thus sustainability continues to be an elusive goal. Whilst individual products proudly proclaim their green credentials, the overall impact of capitalist extractive industrial development continues to accelerate the degradation of the ecological realm. Within critical ecological thought, the very notion of an environmental crisis is a category error (Welsh 2010). The crisis lies in the prevailing social, economic and political systems – as a consequence of the ways humankind conceives of our relationship to the space we inhabit.

Sustainability has been associated with development since the 1987 Brundtland Commission report. This dual role for sustainability, meaning 'ecological care' and 'development' simultaneously, has been critiqued from the beginning. It has been described as the 'conservation of development, not for the conservation of nature' (Sachs 1999, 34). This contradiction has been evident in development discourses since the late 1980s:

> With sustainable development there are no limits to growth. Greens and environmentalists who today still use this concept display ecological illiteracy. There is a basic contradiction between the finiteness of the Earth, with natural self-regulating systems operating within limits, and the expansionary nature of industrial capitalist society. The language of sustainable development helps mask this fundamental contradiction, so that industrial expansion on a global scale can temporarily continue. (Orton 1989, unpaginated)

Sustaining or increasing levels of consumption with more people wanting to consume more resource intensive stuff on a diminishing resource base cannot happen indefinitely. Ecological literacy acknowledges thresholds while also maintaining a critical position on artificially constructed scarcities. With this perspective, it builds capacities to address environmental and social problems where shallow approaches to sustainability fail.

Academics have proposed terms to address the shortcomings in the concept of sustainability. Just sustainability, sustainment, scarcity and degrowth are four concepts (of many) that challenge the hegemony of established sustainability discourses. 'Just sustainability' was coined by environmental educator Julian Agyeman to prioritize justice. 'Sustainment' is a concept proposed by sustainable design theorist Tony Fry as an alternative to the 'defuturing condition of unsustainability' wherein 'myopically, the guiding forces of the status quo continue to sacrifice the future to sustain the excesses of the present' (2009, 1–2). 'Scarcity' reflects the tensions between unlimited human needs and the limited availability of natural resources. But constructed scarcities function to justify austerity measures and punish the majority for the rampant consumption of elites. Constructed or artificial scarcity refers to shortages of resources that are created because of inequitable resource distribution: that is, a constructed scarcity of food (resulting in malnutrition and famine) happens in an African country whose land is used to grow salads for Europeans. Land-grabbing and resource wars create constructed scarcities. The corruption and reckless behaviour of the financial sector results in austerity: a constructed scarcity. The United Nations Food and Agriculture Organization (FAO) and the World Food Program (WFP) are both clear on the fact that there is enough food to provide everyone on the planet with the nutrients they need (UN-FAO 2017; WFP 2017). Malnutrition and starvation are constructed and artificial scarcities, the results of inequitable distribution of resources. The politics of austerity penalize the poor for the extravagant consumption of the rich. 'Degrowth' rejects the de-politicization of sustainability discourses by offering a more precise diagnosis of the problems and prognosis of potential solutions (see Chapter 3, pp. 40–45).

Despite the cynical appropriation of the word 'sustainability' for a variety of unsustainable practices it remains a concept of great consequence, defined as meeting 'the needs of the present without compromising the ability of future generations to meet their own needs' (Brundtland et al. 1987) and it is used in this book. Ecological literacy informs the debate by emphasizing the contextual and collective requirements of sustainability as the condition of an entire culture relative to its gross impact on ecological systems. The per capita ecological footprint of consumption is 4.9 global hectare (gha^1) in the UK and 8.2 gha in the United States (Global Footprint Network 2016) – meaning these nations collectively use respectively between 3.1 and 5.4 times (Leonard 2010, 196) the sustainable level of resources. These two nations have cumulative ways of living that are not sustainable. Globally, people consume the equivalent biocapacity of 1.5 Earths each year (WWF 2014, 9). While many individuals personally use fewer resources and create less pollution, the gross impact of the collective system is the indicator that matters as it is the collective effect that causes total ecological harm. For this reason, sustainability is a political problem about structural choices and not only a matter of individual consumer choices.

The greenwash associated with many uses of the word of sustainability is evident. Since marketing a product or process as sustainable is easier than actually creating sustainable products, and brands have an interest in portraying a green image, the idea of 'sustainability' is used to reassure consumers that conspicuous consumption and business as usual is morally acceptable. Institutions and corporations maintain their legitimacy by publicizing green credentials but are often far less likely to do the much harder work of building capacities to address environmental problems effectively. Ecological literacy enables a more integrated and rigorous analysis than sustainability discourses oriented around green consumption by taking a critical perspective on current development regimes. Within a highly unsustainable world, to be ecologically literate is to be critically informed on the relationships between power, knowledge and ideology that support business as usual – or some slight variation thereof. While some new design approaches are systemic, many continue to lack a critical approach to issues of power. Ultimately, we must confront the cultural practices, development frameworks and interests that reproduce unsustainable development. A critical orientation to issues of sustainability in design is necessary to transform design practice in the context of a deeply unsustainable culture.

Levels and Domains of Learning and Communication

Because the problems associated with sustainability are both very complex and deeply entrenched into our culture, ecological learning involves challenging assumptions and re-examining accepted norms of belief and behaviours. Environmental learning and communication relies on deeper engagement processes than the simple transferral of information. Gregory Bateson described learning levels in the chapter 'The Logical Categories of Learning and Communication' (1972, 279-308) where he developed a framework for learning that distinguishes between levels of abstraction. Sustainability educator Stephen Sterling's interpretation of Bateson's work maps the levels of learning in a trajectory from 'no learning' to deep learning or 'creative-re-visioning' as a four-stage process in environmental education.

Levels of Learning and Communication
No change (no learning: ignorance, denial, tokenism)
Accommodation (1st order learning: adaptation and maintenance)
Reformation (2nd order learning: critically reflective adaptation)
Transformation (3rd order learning: creative re-visioning). (Sterling 2001, 78)

Sterling maintains that learning for sustainability must transcend transmissive learning (1st order learning based on absorbing content) because the transmission of information alone does not necessarily lead to change. He explains 'not only does it not work, but too much environmental information, particularly relating to the various global crises, can be disempowering, without a deeper and broader learning processes taking place' (2001, 19). Instead, environmental learning requires active engagement with new information to nurture both new agencies and new critical capacities.

While most education aims to replicate current worldviews, effective sustainability education must challenge many common assumptions. It therefore requires transformative or epistemic learning – a type of learning that occurs where review of basic premises occurs.

This type of learning entails dialogic processes and engagement in order to help learners process information more deeply. Sterling describes levels of learning in sustainable education:

Levels of Learning in Sustainable Education

A: Education ABOUT Sustainability
- Content and/or skills emphasis
- Easily accommodated into existing system
- Learning about change
- Accommodative response – maintenance

B: Education FOR sustainability
- Additional values emphasis
- Greening of institutions
- Deeper questioning and reform of purpose, policy and practice
- Learning for change
- Reformative response – adaptive

C: Sustainable Education
- Capacity building and action emphasis
- Experiential curriculum
- Institutions as permeable learning communities
- Learning as change
- Transformative response – enactment

(Adapted from Sterling 2001, 2003)

This learning theory holds that concepts move from being beliefs held in the mind to ideas that inform and influence perception, behaviour and attitudes. Once concepts are deeply understood, learners develop new ways of putting ideas into practice in their daily lives (Figure 6.1). The figure below illustrates Sterling's ideas on perception, conception and practice as three parts of the paradigms that structure human understanding and ways of living (2003, 2011).

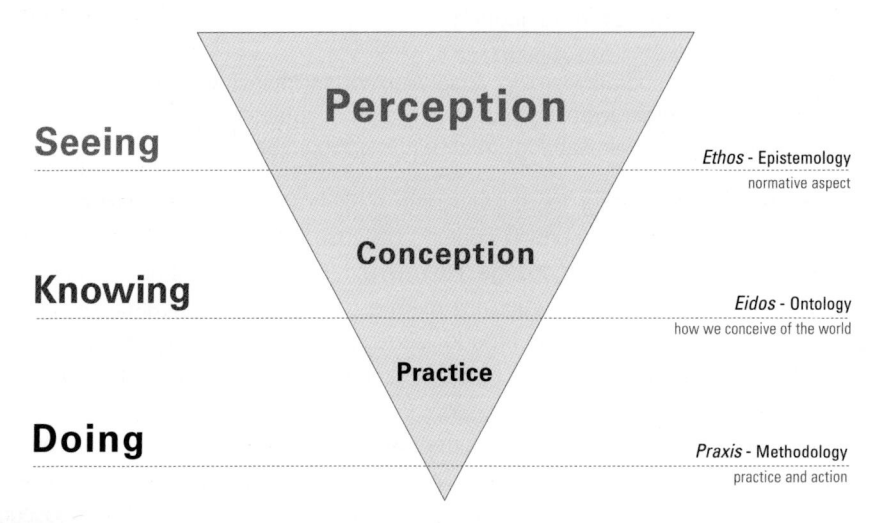

Seeing — Perception — *Ethos* - Epistemology / normative aspect

Knowing — Conception — *Eidos* - Ontology / how we conceive of the world

Doing — Practice — *Praxis* - Methodology / practice and action

Fig. 6.1 *Three Parts of a Paradigm* (following Sterling 2003, 2011). Boehnert, 2017.

The infamous value/action gap in sustainability communication and practice can be understood as the process of learning new concepts in different domains and developing agencies to respond. Sterling's 'Domains of Seeing, Knowing and Doing' theory describes how ideas move from abstract concepts to become a basis of new practices.

Domains of Seeing, Thinking and Doing

1 SEEING – the widening and deepening of our boundaries of concern; the perception and recognition of the broader context in time and space; the concern with context, meaning, and value.
2 KNOWING – the disposition and ability to recognize links and patterns of behaviour and influence between often-disparate forces; to recognize systemic consequences of actions; the concern with dynamics and interrelationships, thinking in terms of flows and patterns rather than distinct entities.
3 DOING – a purposeful disposition and capacity to seek healthy relationships between parts and wholes; recognising that the whole is greater than the sum of its parts; to seek synergies and anticipate systemic consequences of action. (Sterling 2003, 172)

This theoretical model is extended in the Three Parts of a Paradigm model that links each domain with related concepts. The chart below summarizes how paradigms and ideas become manifest, from abstractions to practice, that is, in epistemology, ontology and methodology.

Three Parts of a Paradigm (adapted) (Sterling 2003, 92, 172, 264, 423)

1st	Seeing	Epistemology	Perception
2nd	Knowing	Ontology	Cognition
3rd	Doing	Methodology	Practice

When paradigm shifts occur, changes in thinking and doing need to occur on all three levels. These models explain how it is that ideas can be accepted as intellectual concepts (in the cognitive domain) but not be put into practice. Breaking established patterns of thought and behaviour is not easily achieved. Change requires concerted efforts and strategic interventions. It's possible to intellectually understand the world in a systemic and ecological way, but still perceive it in a reductive manner and continue to act as if one were living in a bubble (Sterling 2003, 122). Participatory action researcher John Heron explains:

> Today, a significant majority of people have abandoned the Newton-Cartesian belief system in favour of some elaboration of a systems worldview. But it may be that they, and certainly the majority of people, still see the world in Newton-Cartesian terms. It is a big shift for concepts to move from being simply beliefs held in the mind to beliefs that inform and transform the very act of perception. (Heron 1992, 251)

This analysis suggests that despite intellectual cultures that no longer operate according to Cartesian logic and reductionism, many of us still see the world in these terms. Supporting ecological theory in principle is not the same as developing the cognitive, perceptual and social capacities to put this awareness into practice. The gap between ideas and actions must to be bridged to make ecologically literacy manifest in everyday life.

Designers are well placed to help negotiate this space from perception to practice. Action emerge from perception and conceptualizations (Figure 6.2) – based on ways of seeing and knowing. Designers use design methods such as mapping to analyse a

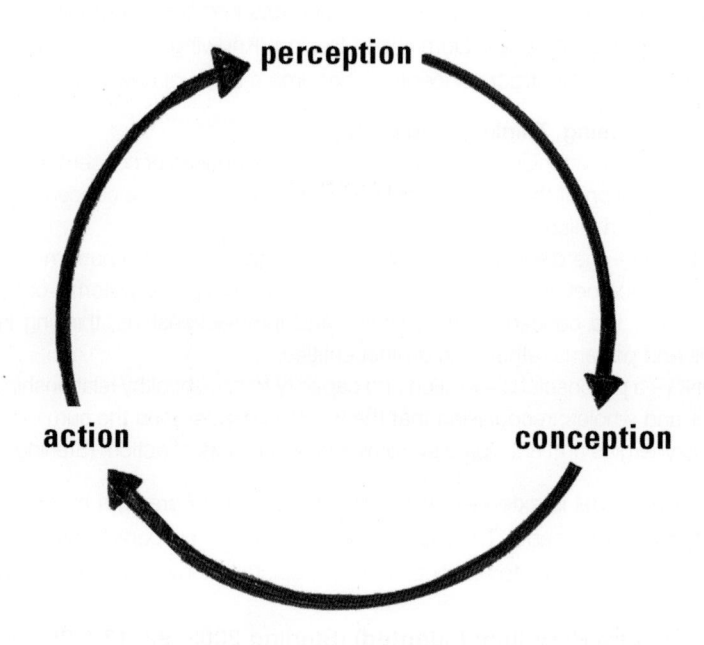

Fig. 6.2 *Perception, Conception, Action*. EcoLabs, 2017.

situation (from multiple analytical perspectives) to inform the design and development of a means to solve a particular problem. The design process creates something new (a visualization, artefact, service, experience, system, code, structure, etc.) that attempts to address an identified problem. This design process then creates a new reality that becomes the new basis for human experience. The new situation affects humans, facilitates certain types of social relations and shapes identities. Design interventions are influenced by human perception and cognition – and then go on to affect humans as an iterative process (Figures 6.2 and 6.3).

Typologies of Ecological Literacy

The daunting task faced by those familiar with the scope of the ecological crisis and the potential for ecological literacy to support effective responses is developing the means of scaling it up to a degree that it is embedded in education and professional practice. Towards this end, ecological literacy can be characterized by two complementary modes. The first (Mode One – Experiential) is typically taught with experiential learning methods in non-formal education and draws insights from philosophy and holistic science. The second (Mode Two – Critical) is more critically engaged and taught through ecopedagogy and transformative learning processes. This second mode is more apt to become institutionalized in formal education due to its links with projects previously established in universities and its critical focus. Mode One provides a conceptual foundation for the second mode, which in

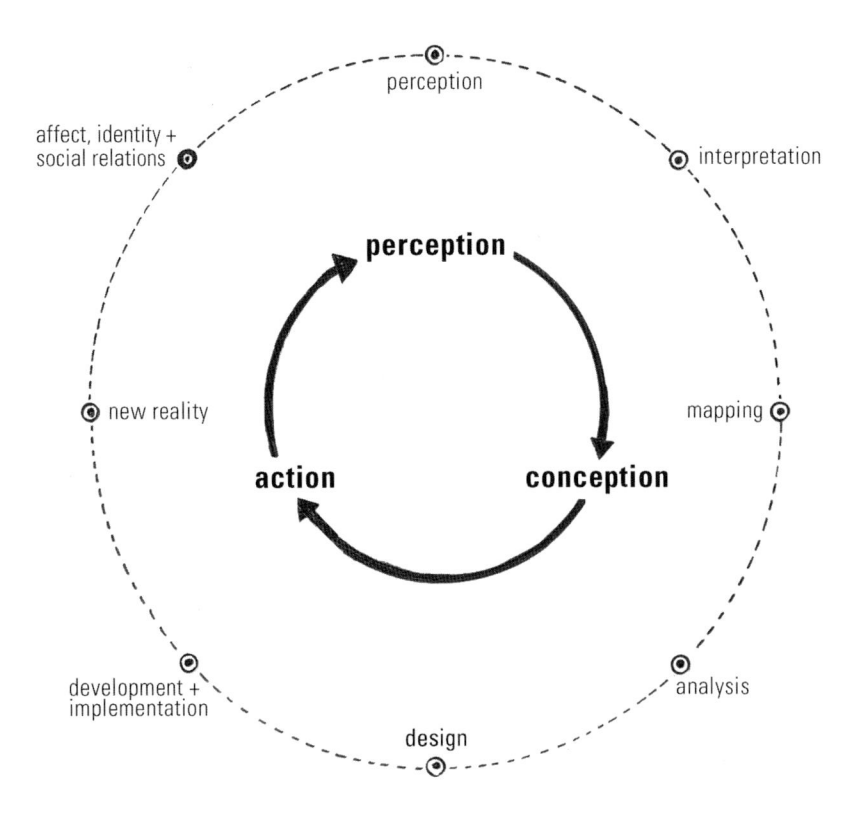

Fig. 6.3 *Design Interventions: The Perception to Practice Cycle.* EcoLabs, 2017.

turn has greater capacities to critique and transform unsustainable institutional practices. Ecological literacy requires questioning underlying premises of theories and methods across disciplines. Mode Two is concerned with putting ecological literacy into practice in various disciplinary practices. It involves bringing the philosophical assumptions of ecological literacy into professions; critical analysis of what is not ecologically literate; and the redesign of structurally unsustainable systems and ways of living.

Mode One – Experiential Ecological Literacy

Mode One is evident outside mainstream education at institutions such as Schumacher College (Totnes, UK) and the Center for Ecological Literacy (CEL) (San Francisco, USA). This mode is characterized by bioregional place-based learning, experiential learning in nature and social learning through dialogical practices. This pedagogy combines developing relationships with place, direct engagement with nature and self-reflexive practice. For both children and adults, Mode One emphasizes experiential learning in nature to support 'a dimension of ecological understanding that has to be subjectively embodied and adapted to the particular local conditions of natural processes in which we participate' (Wahl 2006, 87). The curriculum includes the historical context of emergent ecological theory, holistic science and associated philosophical insights. The arts are seen as a means of accessing ecological subjectivities. Transition initiatives are embraced as community design processes for building resilience. Fritjof Capra, co-founder of CEL in San Francisco, has led the development of this type of ecological literacy through his work with both CEL and Schumacher College in the Totnes, Devon. The weakness of Mode One is that the enthusiasm for consciousness change can be far removed from capacities for transformative actions: actions that always involve critical and political work. The gap between good intentions and critical strategies for interventions on ecologically destructive industrial and political systems is often significant in places where sustainability learning does not link ecological theory to social and political theory, practice and activism. Reflexive and ecologically embedded consciousness must proceed to critical intervention in the systems that perpetuate ecological destruction. Good intentions will not stop ecocidal modes of development. Individuals and communities need critical skills that support agencies to enable transformative work. Critical skills and engagement with the politics of change-making are essential elements of Mode Two, critical ecological literacy.

Mode Two – Critical Ecological Literacy

Over twenty years since its inception ecological literacy remains marginal in many institutional contexts. Facilitating the transition from theory to practice in the context of cultural institutions that reproduce ideas and practices that legitimize and enable unsustainable development is beyond the scope of the relatively apolitical and uncritical Mode One. The work of scaling up and embedding ecological literacy into design education and the wider cultural context requires critical skills, multiple types of literacies, multiple agencies and political engagement. Mode Two supports critical whole systems thinking and new agencies to intervene and transform dysfunctional organizational practices, political systems and power structures. It draws on ecopedagogy and supports critical consciousness. With these practices, Mode Two ecological literacy offers tools and strategies to disrupt and transform the status quo. The fact that so much environmental discourse and education is limited to extremely business-friendly modes of analysis that continue to prioritize profits over ecological priorities is a contributing factor to accelerating ecological harms. Mode Two responds to situations where critical skills are underdeveloped and the roots of the environmental crises are not insufficiently investigated: 'most people consider themselves environmentalists – but commitment is shallow. While environmental education is widespread – it does not challenge entrenched industrial world-view, and imparts only a very thin conception of ecology' (Bower 1999 cited in Dryzek 2005, 200). Since an interest in the idea of 'sustainability' is not always accompanied familiarity with ecological concepts that would enable an integrated analysis of environmental problems or the social theory that explains how unsustainable practice is reproduced, the gap between good intentions and sustainable ways of living is significant. All too often the basic ecological knowledge necessary to inform the design of sustainable systems is absent – despite interest in sustainability as a vague ideal. Mode Two addresses these problems with political engagements as well as change-making practices such as systems thinking.

Ecopedagogy

Ecopedagogy merges ecological theory and critical theory with struggles for social and environmental justice. Ecopedagogy emerged at the Rio Earth Summit 1992 from the tradition of critical pedagogy in the tradition of Paulo Friere. Its socioeconomic critique aims to help 'environmental education move beyond its discursive marginality by joining in solidarity with critical educators' (Kahn 2010, 17) and social movements. Critical theory describes the cultural mechanisms that keep ecological literacy ghettoized. This analysis enables a stronger understanding of the cultural and political forces that must be challenged. Ecopedagogy provides academia with activist tactics and skills to facilitate change institutionally. It provides social movements with critical skills for analysis and strategy development. Critical ecopedagogy is a practice which:

> promotes a dynamic and complex definition of ecoliteracy that seeks to promote the idea that while we are hemmed in by the limits of and interpolated by destructive institutional forms, we can recognize and transcend these thresholds through measures of individual transformational and collective action. (Kahn 2010, 152)

The traditions of critical pedagogy, liberatory education and popular education expand the areas over which learners recognize their own agency to act (on the basis of their own analysis) to challenge conventions and institutions perpetuating oppressions and environmental harms. Central to this empowerment processes is the realization that cultural assumptions arise from historical and cultural circumstances, often as a consequence of ideologies embedded within cultural media, cultural practices and social structures. These ideologies are not always aligned with the values we explicitly support.

Literacies of various types are foundational for ecopedagogy. We are educated by culture, communication, media and design although 'its pedagogy is frequently invisible and subliminal' (Kahn 2010, 73). Multiple literacies are necessary to decipher how culture imposes messages, stereotypes, values and ideologies. Visual, digital, media, data and design literacies create a basis for critiquing images, news, metaphors, myths, cultural stories, digital media, infographics, charts and designed artefacts. Critical thinking helps learners distinguish misinformation and ideologies embedded into cultural content. These literacies enhance capacities for critiquing cultural messaging and building alternatives. Since ecology requires consideration of relationships, a key literacy associated with ecological thought is system thinking. Robust critique and reflexive approaches are a foundation for effective systems work for sustainability.

Systems Thinking

Systems thinking emerged from the critique of reductionist modes of understanding the world and addressing its problems.[2] It takes an expansive, holistic approach that assumes that all the properties of a given system (whether biological, social, economic, mental, linguistic, etc.) cannot be explained by examining the component parts in isolation. Instead, relationships between parts determine how a system as a whole will work. Systems' thinking supports an understanding of context and the underlying patterns that cause events' (Kreutzer 1995, 35). It is an expansive mode of analysis and synthesis that focuses on patterns of relationships, connectedness and context. It refers to a shift in 'focus and attention from things to processes, from static states to dynamics, from "parts" to "wholes" ' (Sterling 2003, 41). Bateson described a way of thinking that looks for 'the pattern that connects' (1980, 8. Systems work nurtures new abilities to perceive patterns and distinguish relationships. Systems thinking emphasizes awareness of cycles, dynamics, patterns including counter-intuitive effects and unintended consequences. It is inherently contextual and relational.

Systems theorist Donella Meadows describes a system as 'an interconnected set of elements that is coherently organized in a way that achieves something ... a system must consist of three kinds of things: elements, interconnectedness, and a function or purpose' (2008, 11). Systems come in different sizes, types and levels of complexity. The concept of an ecosystem, first used in 1935 by botanist and ecologist Arthur Tansley includes living and non-living elements including minerals, gases, earth and water. Flows, feedback, stocks and delays describe a variety of ecological dynamics in systems. Complex adaptive systems possess the ability to organize themselves to deal with new information and respond to feedback. A healthy living system will be both open to feedback and will change in order to preserve itself. Dysfunctional systems fail to adapt to feedback and lose resilience and the capacity to maintain their essential functions. The idea that there are principles that apply to all systems (mechanical, physical, biological, cognitive and social) was introduced by biologist Ludwig von Bertalanfry (1901–1972) over his career and in his book *General Systems Theory* (1968, 32). Cybernetics is the transdisciplinary study of systems. While human constructed systems are ultimately part of nature – only ecological systems have evolved and developed resilience over long periods of time.

Donella Meadows' famous 'Leverage Points: Places to Intervene in a System' (1999) proposes that there is a hierarchy of modes in change-making. Problems associated with sustainability are often located at the more expansive levels of the framework (at the levels of paradigms and norms of behaviour), although more reductive approaches on the top of this list are also useful means to put new priorities into practice.

Places to Intervene in a System (in increasing order of effectiveness)

12 Constants, parameters, numbers (such as subsidies, taxes, standards).
11 The sizes of buffers and other stabilizing stocks, relative to their flows.
10 The structure of material stocks and flows (such as transport networks, population age structures).
9 The lengths of delays, relative to the rate of system change.
8 The strength of negative feedback loops, relative to the impacts they are trying to correct against.
7 The gain around driving positive feedback loops.
6 The structure of information flows (who does and does not have access to information).
5 The rules of the system (such as incentives, punishments, constraints).
4 The power to add, change, evolve or self-organize system structure.
3 The goals of the system.
2 The mindset or paradigm out of which the system – its goals, structure, rules, delays, parameters – arises.
1 The power to transcend paradigms.
(Meadows 1999)

Each layer has an informational effect on the others, so layers can be understood as nested in scope and complexity.

Images are useful tools for working with systems. Visuals facilitate systems thinking by illustrating relationships and dynamics and are routinely used to map systems on various scales. Meadows maintained that discussing systems with words only is inadequate because: 'Words and sentences must, by necessity, come only one at a time in a linear, logical order ... Pictures work for this language better than words, because you can see all the parts of a picture at once' (2008, 5). Since designers are well versed with visual strategies and are typically endowed with spatial intelligence they are in a good position to facilitate systems work. Visual tools such as systems maps, influence diagrams, rich pictures, control models, causal loops diagrams and giga maps enable systems work. Notations describe system processes in short form for graphic displays. Causal loop diagrams display the interrelationships of variables and identify feedback loops in a system. Visual tools are used to illustrate and analyse systems and identify points of leverage. This widespread use of images in systems thinking nurtures relational perception (Chapters 9 and 10) and enhances capacities for sustainable transformations.

System thinking and system sciences are traditions with tensions between instrumentalizing tendencies and reflexive approaches. Systems work has developed in two opposing directions: hard and soft approaches. Hard work (systematic approaches) refers to quantitative processes, engineered for efficiency and optimization. The soft work (systemic approaches) acknowledges the role of the observer; is concerned with connections as both qualitative and quantitative; and considers connecting processes and dynamics as unpredictable. Soft, reflexive systems theory argues that there are limits to what can be known with scientific methods since when an observer studies a phenomenon a critical factor in any analysis is the self-reflective and feedback capacity of the system itself (of which the observer is a part). In the act of observing something, we necessarily change the object under observation and ourselves. This self-reflective dynamic is often absent with systems work that is purely instrumental (systematic). Reflexivity and an awareness of the limitations of predictions and controls are necessary.

Systems, ideas and communication are characterized by different levels of abstraction (Bateson 1972). Epistemological awareness and flexibility supports the ability for 'conscious movement between different levels of abstraction' (Ison 2008, 147). Systems work requires epistemological flexibility that distinguishes between different types of premises, ways of knowing and types of logic appropriate for different types of problems (Sterling 2003; Reason & Bradbury 2006; Ison 2008). For example, the logic that might work to make financial decisions is a different type of logic than what is needed for decision-making involving ecological processes. Soft systems approaches identify appropriate models and methods for different types of problems. This critical and reflexive approach is still marginal. The tradition of radical, second order cybernetics and self-reflexive systems thinkers (especially Gregory Bateson and Francisco Varela) counter more reductive, instrumental and technocratic approaches. Soft systems methods acknowledge both the value and the limits of hard systems approaches.

Systems are a powerful concept. Like all potentially good tools, methods and approaches this power can be deployed in an irresponsible manner and put to nefarious ends. Systems work is not necessarily ecological and it is also used for ecologically destructive purposes.

But ecological thinking always considers systems or networks of relationships (despite reservations by some ecological theorists). Design theorist Jon Goodbun questions if systems as a concept is now 'too contaminated with instrumental associations?' (2011, 222) to be useful since 'these approaches can all too easily become new attempts to achieve control and domination over nature, and/or equally, over humans, when they are unfolded within a capitalist framework' (Ibid., 230). Adam Curtis made a related argument (but more blunt) in *The Use and Abuse of Vegetational Concepts* of the documentaries series on BBC Two (see p. 61). Unfortunately, the documentary reviews an instrumental approach to system works and fails to capture the self-reflective insights of Bateson and others working with soft systems approaches. Clive Hamilton wrote about the dangers of the systems concept when applied to environmental issues:

> Characterising the Earth as a system has unleashed on the planet as a whole a style of thinking – that of the engineer – that had previously been confined to particular elements of it. The Earth as a whole has been opened up to mechanical thinking, updated with the cybernetic ideas of feedback loops, control variables, critical values, system efficiency and so on. (2015, para. 12)

All too often systems work is not ecologically informed, critical or reflexive. Critical system work integrates systems thinking with critical strategies that respond to power and conflict addresses political dynamics (Jackson 1990; Ulrich 1998). Critical and reflexive approaches to systems thinking are foundational for ecological literate design. Ecological literacy should be (but often isn't) a foundation for all systems work.

Ecological Literacy in Design Education

David Orr stresses the importance of ecologically literate design as a means of responding to environmental problems. He explains that environmental problems 'are mostly the result of a miscalculation between human intention and ecological results, which is to say that they are a kind of design failure' (2002, 14). These design failures signal 'inherent problems in our perceptual and mental abilities' but they also suggest that improvements can be made through design (Ibid., 14). For these improvements to be possible, ecological literacy must become a pedagogic priority in design education. Orr describes four prerequisites to ecological literacy:

1. An understanding that 'our health, well-being and ultimately survival depends on working with, not against, natural forces'
2. An understanding of the scope of the crisis and familiarity with 'the vital signs of the planet'
3. A historical understanding of how humankind have become so ecologically destructive
4. An ability to take a practical and participatory approach because 'the study of environmental problems is an exercise in despair unless it is regarded as only a preface to the study, design and implementation of solutions'.

(1992, 93–94)

Ten years later Orr proposed more specific features of ecological design:

1. It is a community process that aims to increase local resilience.
2. It takes time seriously by placing limits on the velocity of materials, transport, money and information.
3. It eliminates the concept of waste and transforms our relation to the material world.
4. It has to do with systems structures. (2002, 180–183)

The struggle to mainstream ecological literacy and embed ecological thinking into professional design practice is situated in the design school. The first challenge for universities is 'to get the entrenched teachers and professors who control the forms of knowledge (including the legitimizing ideology and epistemology) … to recognize the scale and accelerating nature of the ecological crisis' (Bower 2005, 203). Ecologically literate design education is a comprehensive project of learning that requires its own curriculum and research culture in design education. It is not developed in a token green week fashion. Nor is it adequate for sustainable education to be an elective that staff and students can decide to ignore. Ecological literacy requires a dedicated programme that introduces learners to the transdisciplinary challenges of sustainability. Design education must broaden the scope of its inquiry to be relevant to the most important contemporary challenges.

Ecological literacy enables responsible design. Since ecological thought challenges basic assumptions about personal responsibility and boundaries of concern, it is a severe challenge in design education. Sears claimed 'by its very nature, ecology affords a continuing critique of man's [sic] operations within the ecosystem' (1964, 12). Since its scope is so vast, ecological learning requires interdisciplinary collaborations beyond the scope of traditional design education. Since ecological learning presents disturbing information many of us would rather ignore and also disrupts powerful vested interests, it is entirely absent in many places where is badly needed. Despite these difficulties, it is the only viable way to confront environmental problems. We cannot even begin to understand the complexity of our predicament without ecological knowledge and ways of knowing.

Notes

1. Gha = global hectares per person
2. The valid critique of systems theory (in Science and Technology Studies literature and elsewhere) does not negate its usefulness when used in a reflexive and critical way. Systems work expands the scope of analysis and action in design practice.

7   Ecoliterate Design

Ecologically literate design is systems aware, enabling, collaborative and aligned with the patterns and processes of nature. Ecological principles provide time-tested models for the design of sustainable ways of living. Biomimicry pioneer Janine Benyus maintains that good design must use 'an ecological standard to judge the rightness of our innovations. After 3.8 billion years of evolution, nature has learned: What works. What is appropriate. What lasts … [is] based not on what we can extract from the natural world, but what we can learn from it' (1997). While we can't do anything else but design as nature (since we are part of nature): if we want to avoid extinction, we had better 'pay more attention to lessons we can learn from life's evolution' (Wahl 2016, 151). Fritjof Capra describes six 'principles of organisation, common to all living systems, that ecosystems have evolved to sustain the web of life' (2003, 201). The 'Nature's Patterns and Processes' concept developed by Capra and others at the Center for Ecological Literacy (CEL) describes six ecological concepts (networks, cycles, flows, development, nested systems and dynamic balance). I introduce these in this chapter as linked to concepts informing ecological design (resilience, a circular economy, energy literacy, emergence, systems-oriented design, the ecological footprint and planetary boundaries). Since many environmental problems are energy related, energy literacy is part of ecoliteracy. Concepts such as Energy Return On Investment (EROI), the rebound effect and embodied energy are all fundamental for ecologically informed decision-making.

Networks + Resilience

> All living things in an ecosystem are interconnected through networks of relationship. (Stone 2015)

Network science reveals the structures, patterns and organizing dynamics of complex systems by emphasizing relationships. Life organizes itself as networks of relations. Ecosystems are characterized by robust networks with many interconnections. Highly interconnected complex networks such as ecosystems are resilient to shocks and failure because there is a diversity of means for achieving systemic goals. If one node is destroyed, other nodes and links can replace its function.

> Natural systems have a unique ability to survive in a wide range of conditions. Although internal failure can affect their behaviour, they often sustain their basic functions under very high error rates. This is in stark contrast to most products of human design, in which the breakdown of a single component often handicaps the whole device. (Barabasi 2003, 111)

Nature's designs are resilient because networks are filled with redundant nodes. This resiliency is different from design in industrial systems that are typically optimized for maximum efficiency and short-term profitability. Designing for resilience is not the same as designing for efficiency – and very different from design for short-term profitability. Resilient systems can tolerate disturbances and retain their same function and state. They have capacities for self-organization, learning and adaption (Resilience Alliance 2017). The basic design principles of resilient systems consist of small units dispersed in space that are designed at an appropriate scale for redundancy, diversity, decentralized control, quick feedback and self-reliance (Orr 2002, 114–117). Resilient design is flexible, adaptive and enabling. Nature-bridges over motorways are an example of using the networks principle in landscape design for biodiversity conservation that improves bioregional resilience. These are design solutions that enable migrating animals to roam more freely in their habitat now divided up by previously insurmountable obstacles (Wahl 2016, 155). The bridges serve as links in a network that enable animals to survive.

In the Transition Town movement, local resilience is developed through community design processes (see Chapter 9, p. 105). Transition groups work towards the co-construction of a 'local energy descent plan' to adapt to climate change in the wake of the failure of national and international governments to adequately respond. Here the bioregion is a central organizing principle towards building resilience into local spaces. Sustainable design writer John Thackara describes the relevance of this focus:

> A bioregion re-connects us with living systems, and each other, through the places where we live. It acknowledges that we live among watersheds, foodsheds, fibersheds, and food systems – not just in cities, towns, or 'the countryside' … Growth, in a bioregion, is redefined as improvements to the health and carrying capacity of the land, and the resilience of communities. Its core value is stewardship, not extraction, a bioregion therefore frames the next economy, not the dying one we have now. (Thackara 2015b)

Transition initiatives work for an economy based 'stewardship, not extraction. Growth, in this new story, means soils, biodiversity and watersheds getting healthier, and communities more resilient' (Ibid.). The movement develops project such as local currencies, local food growing networks, repair shops, community owned energy infrastructure and other examples of locally owned sharing and solidarity economy enterprises as a basis for local resilience. Social learning and network building processes are part of the transition, resilience development process.

Cycles + A Circular Economy

> Members of an ecological community depend on the exchange of resources in continual cycles. (Stone 2015)

Cycles are an obvious pattern in nature (i.e. days, months, years, water cycle, carbon cycle, etc.). Within ecological systems all elements are endlessly re-used. In stark contrast to these cycles, the economic system currently depends on a continuous linear flow of natural resources, extracted from the Earth and then moving through industrial production, consumption and disposal processes, resulting in various types of pollution. Economic growth has material demands and thus the need for resources and energy continues to grow as does pollution (e.g. climate change, toxins in the food chain) and resource depletion (water scarcity, minerals depletion, etc.). Since 99 per cent of the materials extracted from the Earth becomes 'waste' in just six months, industrial systems are highly wasteful[1] (Lovins, Lovins & Hawkins 2007/1999, 81). Although this is an old statistic and some products have become less resources intensive since, net waste continues to rise. Americans make 4.6 pounds of municipal solid waste per person per day (Leonard 2010, 244) and globally 3.5 million tons per day (Hoornweg, Bhada-Tata & Kennedy 2013). In response to this dilemma, advocates of a circular economy aim to eliminate the concept of waste by applying nature's cycling processes to industrial ecology and design. The Royal Society of Arts developed a framework for the circular economy with its 'Four Design Models for Circularity' consisting of design for longevity; design for leasing or service; design for re-use in manufacture; and design for material recovery (Figure 7.1). The circular economy imitates 'nature's highly effective cradle-to-cradle system of nutrient flow and metabolism in which the very concept of waste does not exist' (Braungart & McDonough 2002, 103–104; RSA Action and Research Centre 2013). When incorporated into a larger vision of economic, political, social and cultural transformation, the circular economy offers transformative potential. To address problems on a scale that will make a difference, the model of development itself must be designed to support a circular economy, reflecting the economies in the natural world. A circular economy must be deployed all scales, micro to macro such that the entire economic system reflects circular principles, working with, rather than against, natural forces.

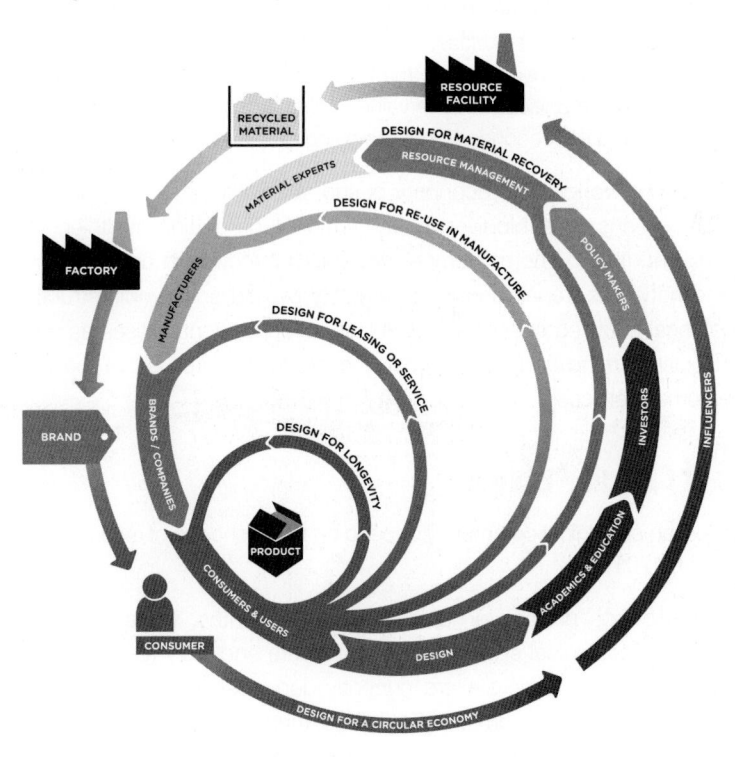

Fig. 7.1 *The Four Design Models for Circularity.* RSA.

Flows + Energy Literacy

Each organism needs a continual flow of energy to stay alive. The constant flow of energy from the sun to Earth sustains life and drives most ecological cycles. (Stone 2015)

Flows of energy and other nutrients provide organisms and ecosystems with what they need to flourish. The availability and flow of energy and other natural resources is also essential for the economic system. Organisms, ecosystems and the economy are all dependent on energy. In the context of increasing resource scarcity and climate change, issues of the flow of energy and natural resources are increasingly complex and precarious. The complexity of global energy flows is illustrated in Figure 7.2, which uses a Sankey diagram to display the flow of various types of energy from fuel to service.

Energy literacy refers to an awareness of the benefits and consequences of various types of energy; the risks associated with energy extraction and climate change; the scale of the change necessary to respond to climate change; and other issues associated with all of the above. Energy literacy includes carbon capability which transforms the 'understanding of carbon from an inevitable waste product of modern lifestyles to a scarce and potent resource that needs to be carefully managed' (Whitmarsh et al. 2009, 126). *The Handbook of Sustainability Literacy* (2009) lists learning objectives associated with carbon capability as an awareness of:

- the causes and consequences of carbon emissions;
- the role individuals – and particular activities – play in producing carbon emissions;
- the scope for (and benefits of) adopting a low-carbon lifestyle;
- what is possible through individual action;

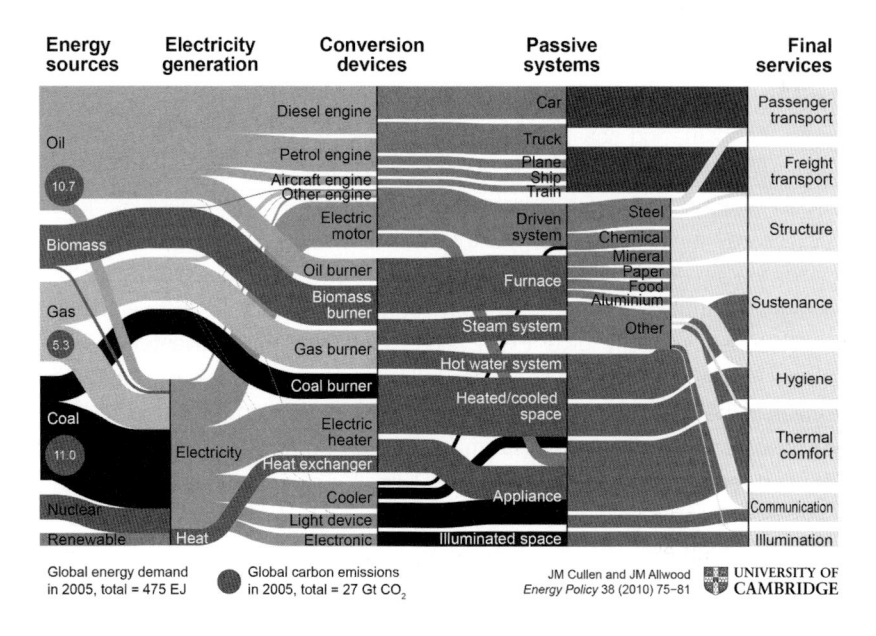

Fig. 7.2 *Tracing the Global Flow of Energy from Fuel to Service.* Image derived from research in: Cullen and Allwood (2010).

Ecoliterate Design 91

- carbon-reduction activities which require collective action and infrastructural change;
- managing a carbon budget; – information sources – and the reliability (bias, agenda, uncertainty, etc.) of different information sources; and
- the broader structural limits to and opportunities for sustainable consumption. (Whitmarsh et al. 2009, 126)

Economic development is energy intensive and all developed nations are currently reliant on fossil fuels. Energy security and energy justice have become competing discourses. Energy security highlights artificially constructed scarcities (see pp. 90–93) and describes dependency on fossil fuels as a critical vulnerability. Energy justice is a counter-discourse that aims to redistribute access to energy. Resources such as David Mackay's *Sustainable Energy – Without the Hot Air* (2009) explore various energy options in response to climate change. The Centre for Alternative Technologies' *Zero Carbon* (2010, 2015) reports map pathways to de-carbonize the economy in the UK using currently available technology. These resources provide insights on how personal, social, technological and political change can be enacted to dramatically reduce carbon emissions. While knowledge of personal responses to energy problems is necessary, since many of the solutions to energy related problems can only be enacted at the state or international level, energy infrastructure is a political and structural problem. Energy return on investment, the rebound effect and embodied energy are analytical concepts for making informed design decisions on energy issues.

Energy Return On Investment

Industrial development and modern ways of living have been made possible by relatively cheap energy. Ecology professor Charles Halls explains, 'You can't have an economy without energy. Energy does the work!' (quoted by Thackara 2015, 12). One barrel of crude oil contains, in energy terms, the equivalent of the heavy manual labour of twelve people working for one year (Wijkman & Rockstrom 2011, 62). As easily accessible fossil fuel supplies diminish, the era of cheap fossil fuel energy is ending. The concept of 'peak oil' refers to the increasing scarcity of *easy to access and relatively cheap* conventional fossil fuels. One way to understand the consequences of energy scarcity is by measuring EROI, that is, 'Energy Return On Investment'. EROI is the ratio of energy acquired from a particular energy resource to energy expended to obtain that energy resource. Lower numbers indicate higher production and extraction costs. The fact that historically the 'EROI of our most important fuels is declining and most renewable and non-conventional energy alternatives have substantially lower EROI values than traditional conventional fossil fuels' (Hall, Lambert & Balogh 2014) has had profound impacts on the economy – but things are changing fast. In the early and mid-twentieth century, oil production provided 50–100 times the energy invested (Wijkman & Rockstrom 2012, 60). Today, energy from unconventional fossil fuels and renewables has much lower EROIs. For example, the Tar Sands have an EROI of between 3 and 5 (Ibid., 60). Shale gas (procured by fracking) has an EROI of 5–7. Renewable and green electricity is getting more efficient and price effective. Reports that 'the energy return on investment (EROI) of solar has overtaken that of oil … heralds the dawn of the age of solar' (Bond 2017). Hydroelectric as an EROI of 40+, wind about 20, and solar between 6 and 22 (Carbon Brief 2013; Inman 2013; Bond 2017). The EROI nuclear energy is highly disputed with independent studies showing an EROI of 5 (Lenzen 2008) and the World Nuclear Association estimating forty to sixty (Carbon Brief 2013). Thackara explains that it is not so much an energy crisis that we face, but 'an *exergy* crisis – that is a shortage of energy that is so highly concentrated, and easy to obtain, that it can be used to drive the economy' (2015a, 11). This energy transition is profoundly disruptive and challenging because the 'foundation myths of the modern age – reason, progress, mastery over nature – are all oil-powered narratives. In the 1950s, when Milton Friedman expounded the economic thinking that dominates political discourse to this day, you could buy a barrel of oil for $3.50' (Ibid., 12). The era of cheap fossil fuel energy is over and the sooner we can build new solar, wind and water energy infrastructure the better.

The Rebound Effect

The rebound effect (also known as the Jevons paradox) refers to the phenomenon wherein savings through energy efficiency are spent on other goods and services (such that gross consumption of energy grows rather than contracts, despite work done implementing more efficient systems). In the landmark report *Sustainability Without Growth?* former Economics Commissioner of the UK Sustainable Development Commission Tim Jackson describes a situation where an individual spends

> the savings from energy-efficient lighting (say) on a cheap short-haul flight ... This somewhat counter-intuitive dynamic helps explain why simplistic appeals to efficiency will never be sufficient to achieve the levels of decoupling required for sustainability. (2009, 62)

Clearly, it is the gross impact of carbon in the atmosphere that matters in terms of climate impacts. So if the gains in efficiency are dwarfed by increases in consumption and the rebound effect results in more greenhouse gas pollution – we are not effectively addressing climate change with these gains in efficiency. Considering that over the past three decades, despite a 30 per cent increase in resource efficiency, global resources use has expanded 50 per cent (Flavin 2010, xvii) new consumers, expanded markets and the rebound effect are driving harms associated with resource scarcity and pollution. Jackson sums up this problem: 'In short, efficiency hasn't outrun scale and shows no signs of doing so' (2009, 76). Design theorist Cameron Tonkinwise describes what the rebound effect means for design:

> The rebound effect clearly points to the futility of built environment change without lifestyle, habit and expectation alteration. Far from excusing social change, technological or built environment change needs social change for its contributions to sustainability to be sustainable. Sustainable technologies are only as sustainable as the lifestyles that take them up. There are no sustainable products or buildings, only ever the sustainable use of products or buildings. The slogans used by government and non-government sustainability advocates in the late nineties, such as 'change your light bulb, not your lifestyle,' are utterly misdirected. There can be no built environment change without lifestyle modification, and there can be no sustainable built environments without lifestyles modified to be more sustainable within those built environments. (2007, 8)

The rebound effect explains why technocratic approaches (emphasizing efficiency) or the responsibilization discourse (emphasizing consumer choices) are ineffective without changes in system structures that encourage or even determine energy use.

Embodied Energy

The embodied energy of a product has to do with the energy that is required by all activities in production and disposal processes, including resource extraction, processing, transportation, production, sales, installation, disassembly, deconstruction and/or decomposition (i.e. safe disposal). Typically most of this energy has already been used by the time the customer acquires the product. Using life cycle analysis, embodied energy should include the energy required for repair, maintenance and disposal or reuse. Potentially embodied energy should include a share of the embodied energy in tools and other instruments that are required to service these processes. Obviously, embodied energy is complicated to calculate. Nevertheless, it is the only way to make realistic judgements of the energy merits of different energy policy proposals and design options. For example, the fuel that a car uses or the electricity that a washing machine uses is only part of their overall energy impact due to the embodied energy in production processes. For this reason, it does not always make ecological sense to replace old appliances, cars, etc. with more energy efficient models.

Embodied energy dramatically challenges simplistic perspectives on issues of energy and climate change. For example, Jackson describes a situation where progress on UK's Kyoto targets was undermined by a failure to account for embodied energy: 'An apparent reduction in emissions of 6 per cent between 1990 and 2004, as reported under UN FCCC guidelines is turned into an 11 per cent increase in emissions, once emissions embodied in trade are taken into account' (2009, 51). While the UK likes to take credit for lower greenhouse gas emission in this time period, what has happened is that production processes moved abroad while UK rates of consumption of carbon intensive products increased. The map here (Figure 7.3) displays emissions associated with consumption in different nations.

The fact that energy is embodied in products that are transported around the world means that soaring carbon emissions in China are linked to globalization of trade and increases in consumption in North America and Europe. For decades politicians worked with the World Trade Organization (WTO) to create trade agreements that enabled production processes to move to where labour was cheapest. These international laws were deeply controversial with trade unions and the alter-globalization movement and sparked 'The Battle of Seattle' WTO

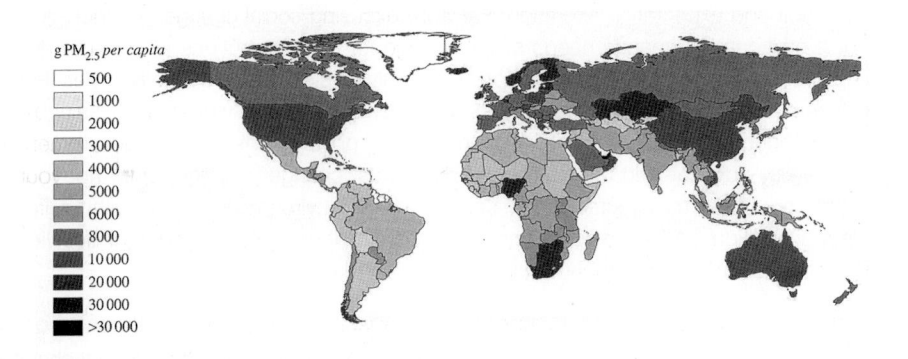

Fig. 7.3 Per capita consumption-based emissions, Meng et al. 2016.

protests in Seattle in 1999 (amongst other sites of contestation over the past thirty years). The laws went ahead despite fierce opposition. Industry moved from America and Europe to China and other places where labour is inexpensive. The steep increases in Chinese carbon emissions over the past decade are a result of this globalization of trade dynamic. Products are produced in China and other lower-income nations but consumed in North America and Europe. China has high emission but (1) many Chinese people work in factories making inexpensive items for the richer nations; (2) per capita emission rate in China are lower than in North American (although they are equivalent now to Europe); and (3) a large portion of the per capita emissions in China is embodied energy in products that are enjoyed elsewhere. While the carbon intensity of electricity is higher in China than Europe (i.e. they mainly burn coal to get electric rather than cleaner sources) and this is certainly a problem, since people elsewhere are enjoying goods made in China that have resulted in the dramatic increases in China's emissions, it is hypocritical for Europeans and especially North Americans to complain about China's greenhouse gas emissions. The following two figures display: (a) the flows of emissions embodied in trade from net exporting countries to importing countries and (b) the flow of emissions between regions (Figure 7.4).

A familiarity with embodied energy reveals just how complex energy issues are when considering transnational industrial processes and trade, how hard it is to calculate responsibility and also how those with more power control the discourse. This confusion can obfuscate responsibilities and issues of energy justice. Concepts such as EROI, rebound effect and embodied energy are basic building blocks for informed decision-making in design and the development of energy policy concerned with issues of justice. Energy democracy, energy justice and the prospect of energy descent will become increasingly important as we adapt and mitigate to climate change. Since energy literacy is important not only in design but in society at large, designers also have a role to play in promoting energy literacy by making greenhouse gas emissions tangible and energy reduction strategies accessible. Designers can help with strategies to help budget carbon (such as carbon calculators and monitors), support energy literacies and with strategies for energy descent. In the image 'My 2007 Diet', Saul Griffith makes the energy in his diet visible and links person energy use and climate change (Figure 7.5).

Development + Emergence

> All life – from individual organisms to species to ecosystems – changes over time. Individuals develop and learn, species adapt and evolve, and organisms in ecosystems coevolve. (Stone 2015)

As complex adaptive systems develop they exhibit self-organizing properties. Development is a learning process in which 'individuals and environments adapt to one another' (Capra 2005, 27). Emergence is a process of self-organization of complex adaptive dynamic systems that results in the creation of entirely new properties. Emergence appears as the outcome of relationships wherein the whole becomes greater than the parts. The phenomenon of emergence is significant for sustainability because it describes how systems exhibit unpredictable behaviour. Emergent properties can have positive or negative implications, but it is impossible to predict outcomes with certainty. Reductionist science tends to ignore emergence and this dismal of the unexpected contributes to illusions of the prediction and control – with

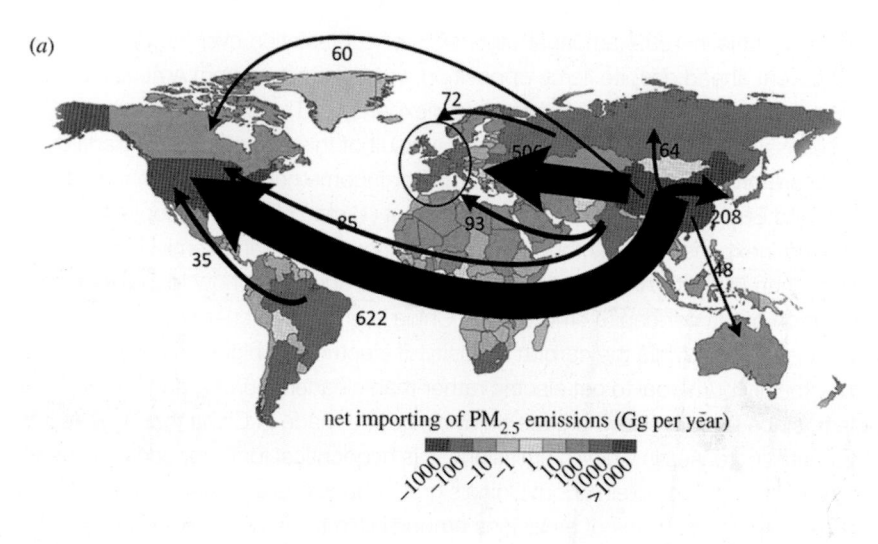

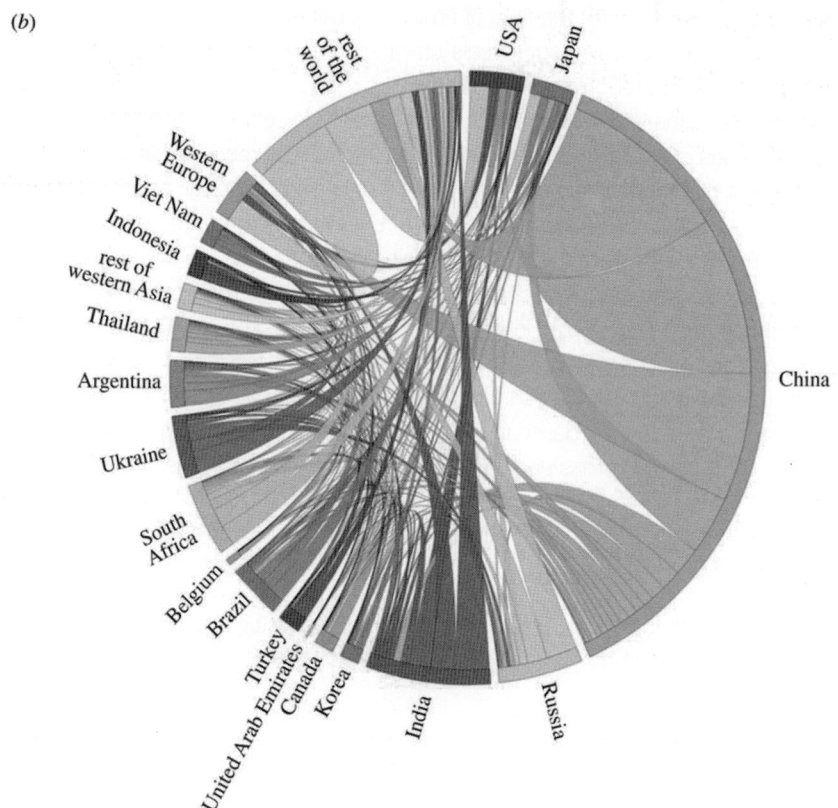

Fig. 7.4 *Production-based emissions linked to foreign consumption*. (a) The largest interregional flow of emissions embodied in trade from dominant net exporting countries to the dominant importing countries. (b) The interregional flow pattern of emissions embodied in the global trade system (Meng et al. 2016).

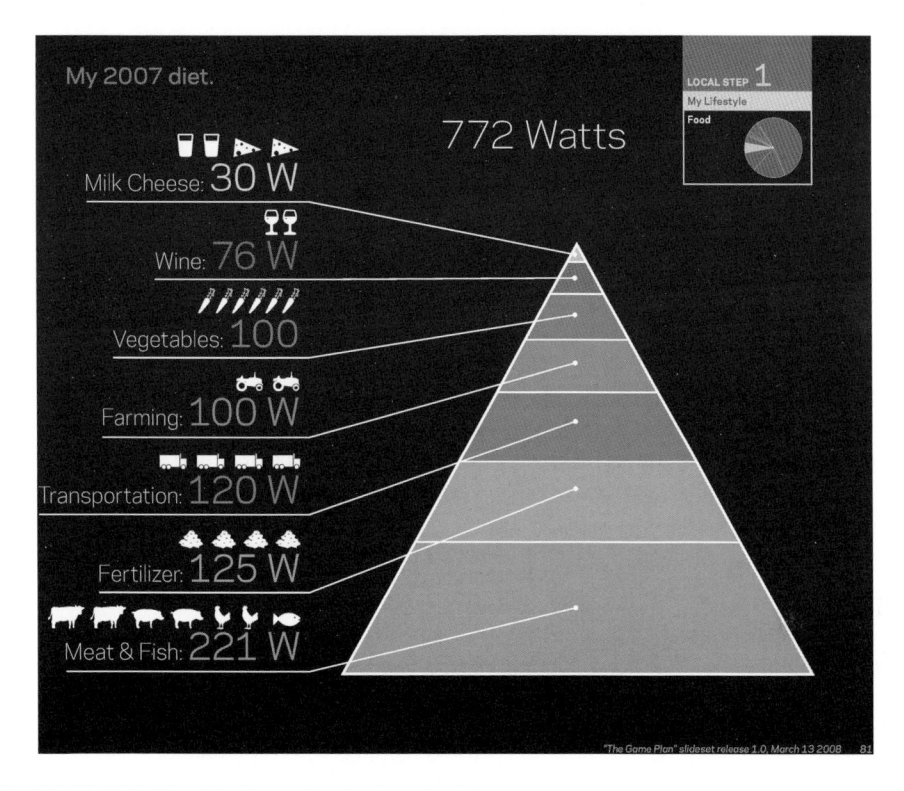

Fig. 7.5 *My 2007 Diet, The Game Plan* by Saul Griffith, 2008.

the ignoring of the so-called unintended consequences[2] until after sometimes irreparable ecological damage is done.

Emergence is a dynamic of change in all living systems. Living systems constantly seek self-renewal (Capra 2005, 27) and have the ability to create life: 'All living beings create themselves and then use that "self" to filter new information and co-create their worlds' (Wheatley 2006, 167). Information generated by feedback processes is used to generate new capacities. Change occurs 'only when an organism decides that changing is the only way to maintain itself' (Ibid., 20). Healthy living systems use new information to preserve themselves. Dysfunction and disease occur when living systems ignore information that threatens their capacities to survive over time and fail to respond to new threats.

Ecologically literacy enhances society's collective capacities to attend to sustainability challenges by encouraging our own self-organizing activities (such as the design of sustainable ways of living). Increasing relational, systemic and critical thinking are emergent processes of reflexive self-organization as humankind organizes to respond to complex problems. Ecological literacy itself can be understood as an emergent phenomenon. Reflective and systemic ecological awareness supports self-organizing capacities that enable the design of more sustainable futures. Strategies and tools developed with this mindset can make it possible to design response strategies that take emergence into account. A helpful communication design strategy is the process of mapping unknowns. Here phenomena that are not yet known are made visible in a mapping process intended to reveal possible unintended consequences before, rather than after, damage has been done. But emergence remains unpredictable. Since reductionist and instrumental approaches to sustainability ignore emergence, these modes have limited capacity to address environmental problems.

Wicked problems are characterized by outcomes that are unpredictable. Wicked problems are: 'a class of social system problems which are ill-formulated, where the information is confusing, where there are many clients and decision-makers with conflicting values, and where the ramifications in the whole system are thoroughly confusing' (C.W. Churchman quoted in Buchanan 1992, 15). The concept of wicked problems highlights indeterminacy. Wicked problems stress the difficulty of formulating an exact problem and arriving at a definite solution. These problems call for interdisciplinary collaboration because 'individual intelligence is insufficient to our tasks' (Rittel & Webber 1973, 160). The unpredictability of complex systems demands working with the precautionary principle and seeking solutions beyond reductionist and mechanical ones. When dealing with wicked problems, precaution must become an operating principle.

Nested Systems + Systems-Oriented Design

> Nature is made up of systems that are nested within systems. Each individual system is an integrated whole and – at the same time – part of larger systems. (Stone 2015).

Ecological systems are nested systems. Problems arise when the relationships in nested systems break down. Ecological economists describe unsustainable development as caused by a dysfunctional relationship between the economic, social and ecological systems. The economic system currently does not operate as a subsystem of the larger ecological system in which it is embedded and ignores information vital for its survival over time. The implications of dysfunction in nested systems can be dramatic. A subsystem can be parasitic when it does not acknowledge itself as interdependent with its context (see pp. 70–71). Systems-oriented design is an expansive approach to design that uses systems thinking (and its awareness of nested systems) to address problems that straddle environment-social-economic-political spheres (see pp. 121–123). *The Steady State Economy* (see Figure 7.6) illustrates levels of nested systems in the ecological economist Herman Daly's conceptualization of a steady-state economic system. The image also illustrates different levels of abstraction from the theoretical on the top to concrete policy measures and practices on the bottom.

Dynamic Balance + the Ecological Footprint

> Ecological communities act as feedback loops, so that the community maintains a relatively steady state that also has continual fluctuations. This dynamic balance provides resiliency in the face of ecosystem change. (Stone 2015)

Dynamic balance is created as ecological systems organize themselves in response to feedback from subsystems and meta-systems. Ecological systems maintain their processes through feedback loops that allow nested systems to self-regulate within tolerance limits (Capra 2005, 28). These limits can be described in various ways using ecological assessment

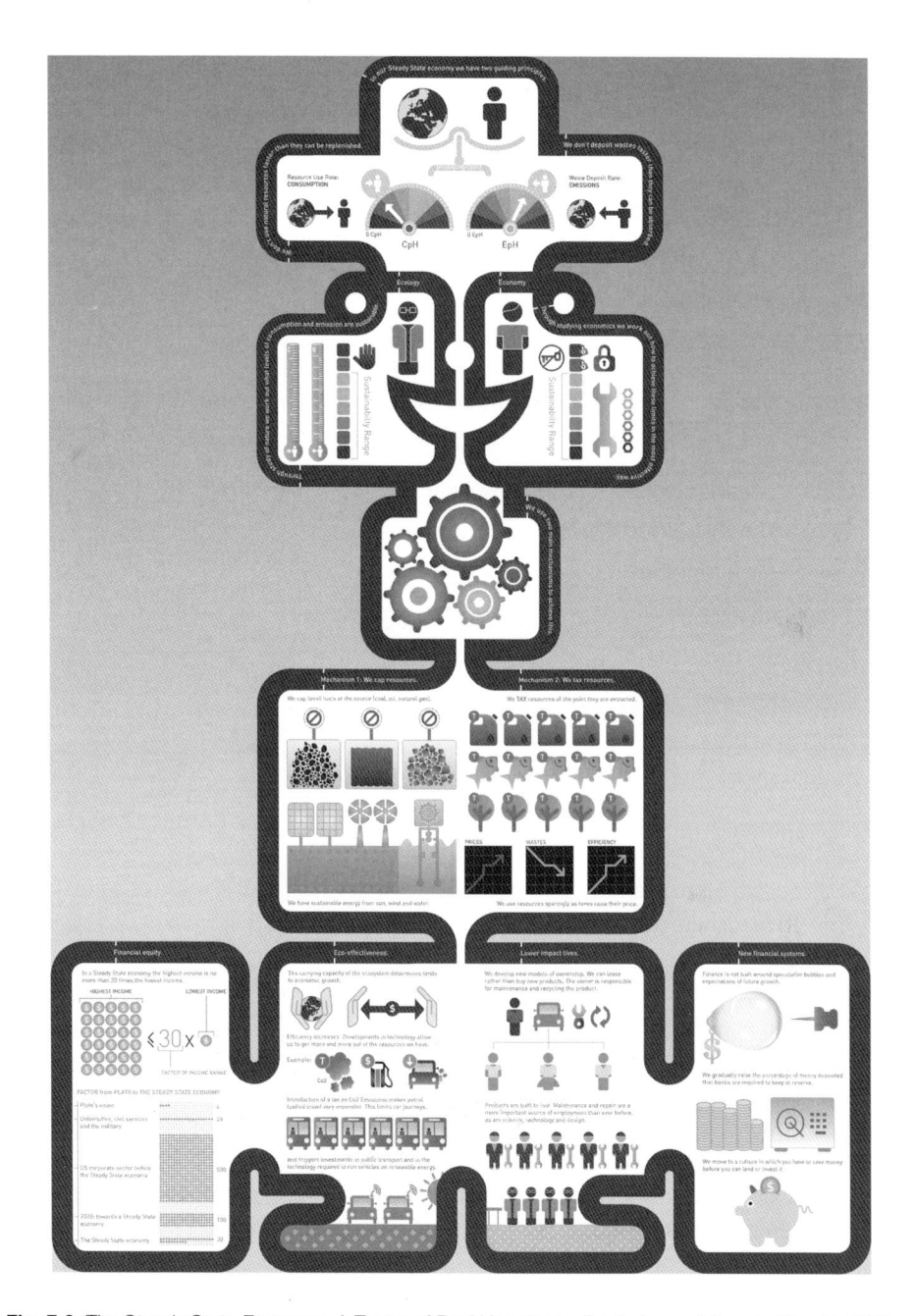

Fig. 7.6 *The Steady State Economy: A Totem of Real Happiness.* EcoLabs and Angela Morelli, 2009.

tools such as the ecological footprint, a concept first conceived by Mathis Wackernagel and William Rees at the University of British Columbia in the early 1990s. The ecological footprint is a metric that calculates human pressure on the planet by measuring how much 'land and water area a human population requires to produce the resources it consumes and to absorb its carbon dioxide emissions, using prevailing technology' (Global Footprint Network

2017). Ecological accounting tools determine the area of productive land required for populations to maintain current consumption patterns. Tolerance levels are determined by how much stress an ecological system is under due to resource extraction, pollution and other human activities. If ecosystems are damaged beyond critical thresholds, dramatic change and even collapse can occur on various scales.

Another concept that describes ecological thresholds is the Planetary Boundaries framework. It establishes tolerance limits of various Earth Systems and a 'safe operating space for humanity' (Rockström et al. 2011). The nine Earth-System Processes in the Planetary Boundaries framework are listed below.

Planetary Boundaries Framework (Stockholm Resilience Centre 2017)

1. Climate change *
2. Biodiversity loss and extinctions/Biosphere integrity *
3. Biogeochemical cycles (Nitrogen + Phosphorus) *
4. Land use/land-system change *
5. Ocean acidification
6. Freshwater use
7. Ozone depletion
8. Atmospheric aerosols
9. Chemical pollution/Novel entities

*Boundary crossed beyond 'safe operating space for humanity'.

Four of these boundaries have already been severely breached: climate change, biodiversity loss, biogeochemical cycles and land-system change. Land-system change is in the zone of uncertainty, where it is in increasing risk as it has already crossed a safe boundary. Transgressing planetary boundaries presents serious risks of dramatic ecological destabilization including ecosystem collapse on various scales. Planetary boundaries are receiving widespread attention within scientific communities but are still far from being integrated into the design disciplines. Design must respond by creating ways of living that ends the assault on planetary boundaries.

Making ecological principles explicit and visible supports ecological learning. Networks, nested systems, cycles, flows, development and dynamic balance can be visualized. Manuel Lima's work on networks illustrates patterns in nature and in human constructed systems (2011). Nested systems are visible in biology and geography textbooks with cell structures and with detailed views of geographic maps. Cycles are seen in visual representation of water cycles and carbon cycles. The flow of materials and energy can be illustrated as it moves through economic systems with flow diagrams and Sankey charts. Ernst Haeckel illustrated the concept of development in series embryo drawings. Dynamic balance is visualized in the images of the Planetary Boundaries framework (Figure I.3). These visualizations support ecological learning.

The design methods I describe in this chapter are united by expansive approaches to the interconnectivity and complexity of nature's processes and patterns. With this perspective, it is evident that sustainability 'is not an individual property, but a property of an entire network' (Capra 2005, 23). Since it is the collective impact on the natural commons that will determine sustainable ways of living now and future conditions, viable futures can only be achieved through systemic understanding and collaboration between all elements of a network.

Notes

1. Most of this waste is the by-product of production and shipping processes.
2. Some consequences can be anticipated but are widely ignored. For example, we know about the damages of carbon dioxide in the atmosphere but we have not created policies to stop the activities that contribute to this problem.

8   Ecological Movements

Ultimately sustainability is not a feature of a particular product or service but the collective condition of a culture relative to its impact on ecological systems. Since the cumulative impact of modern ways of living are destabilizing the climate; causing record-high rates of species extinctions; depleting fresh water supplies, forests, fertile soil, glaciers and ice at the poles; creating toxic lakes and underground water reservoirs, air pollution, radioactive wastes, sludge dumps and garbage patches in the oceans; and leaving irreparable devastation on the sites of mountaintop removals and tar sand – amongst other ecological harms: this civilization is not sustainable. While the behaviour of certain individuals is below the threshold,[1] the gross impact of all of humanity on Earth Systems is the indicator that matters. Ecological literacy emphasizes the contextual, relational and collective characteristics of ecological well-being as central to the pursuit of sustainability. Since the very concept of collective responsibility for environmental harms is rejected most conservatives and neoliberals, the environmental crisis presents a severe philosophical and political problem in the current political climate.

Environmental social movements respond to the failure of social institutions and governments to provide adequate protection for the environment and people who rely on it. In this chapter, I briefly review major movements pushing forward environmental justice agendas and describe some of the ideas and practices that are generated by these movements. Theories and traditions such as deep ecology, permaculture, anarchism, eco-Marxism, agroecology, traditional ecological knowledge and ecofeminism all make critiques of mainstream environmentalism and offer alternatives that can inform design. Social movements[2] and NGOs[3] campaign against environmentally destructive policies and practices while simultaneously building sustainable alternatives. In many cases, those who become active in these movements are regular people responding to local development projects that threaten to destroy environmental spaces leaving long-term public health issues and depleted, degraded and/or polluted environmental spaces in exchange for a few temporary jobs. The pragmatic focus in social movements on needs, values and well-being exposes structural problems within capitalism that makes sustainability so hard to achieve within the current economic paradigm.

Ecological Movements as Micro-Experiments in Transition

Mainstream environmental discourses hold that environmental problems can be addressed within the current economic system with market mechanisms, with clean energy and other innovative solutions. Governments, social institutions and corporations attempt to respond to environmental problems using these approaches. Meanwhile, social movements respond to environmental problems and associated social problems with a plethora of activities to address the concerns of front-line communities, expose misinformation campaigns supported by corporate interests (the climate denial industry, the corporate agriculture lobby, extractive industries, etc.), stop new polluting or ecologically destructive infrastructure and build alternatives. In the following section, I review various environmental movements and describe the insights and strategies for ecologically viable alternatives that they offer. These movements can be understood as experimental laboratories offering visions and practical efforts towards constructing new ways of relating to one another and non-human nature. Many of these movements are explicitly working towards post-capitalists alternatives for reasons I describe below.

Environmental Justice

The term 'environmental racism' was coined in the late twentieth century with research that exposed how race and environmental injustices intersect. Robert Bullard demonstrated that black Americans disproportionately have waste treatment facilitates located in their neighbourhoods (Bullard 1990). Environmental justice movements see environmental problems as a result of a system that reproduces injustices where the greatest impact is felt by those least responsible for pollution. Since 80 per cent of resources are consumed by just 20 per cent of the Earth's population (STWR 2014, 18) and the negative environmental impacts of this consumption are concentrated on the poor (e.g. landfills, incinerators, toxic operations, mining, resources extraction, are all more likely to be located where the poor live), environmental problems are also problems of justice – in this case environmental justice. Since the poorest 20 per cent are left with only 1.3 per cent of global resources (Ibid., 18), the distribution of natural resources is also an issue of economic justice.

Despite these facts, mainstream environmental discourses often identify overpopulation (and implicitly poor as those who are having too many children) as a central cause of the environmental dilemma. This focus on overpopulation has been an ongoing site of dispute since Thomas Malthus (1766–1864) published *An Essay on the Principles of Population* (1798). Social justice advocates have disputed this proposal for over 200 years. The basic problem with the population argument is that it neglects gross disparities in consumption of resources. The average American uses over ten times the amount of natural resources as the average person in India, Haiti or Eritrea (WWF 2014, 36–37). Obviously within America consumption is far from equitable as the wealthy use many hundreds of times more resources than the poor. Justice advocates argue that, within the context of vastly different patterns of resource use, it is clear that overconsumption, inequity and structural dynamics of the economic system – not population – are the primary drivers of environmental problems.

Environmental justice movements seek to reduce pollution and other environmental harms while also addressing issues of social justice and equity. They are typically sceptical about claims that the free market can solve environmental problems. The radical positions hold that capitalism is the primary factor driving climate change and other environmental harms since capitalism always prioritizes capital accumulation[4] over other priorities. Accordingly, new ways of organizing social relations and the political economy must be created to respond to issues of social and environmental justice. Deep ecology, permaculture, anarchist, eco-Marxism, agroecology, indigeneity and ecofeminism all challenge mainstream environmentalism. While there are long–standing disputes, there are also important commonalities. They all hold ontologies that highly value the environment; they all problematize hierarchies (in different ways); they all respond critically to the logic of domination; they promote a systemic or holistic view of human–nature relations (although some reject the term 'holistic'); and they all subscribe to a commons-based approach that is sceptical of market-based mechanisms and corporate environmentalism. The theories and practices developed by these groups map alternative ways of relating to the natural world.

Deep Ecology

Deep ecology refers to a deep questioning of all practices, values and assumptions that result in environmental harm, as distinct from shallow environmentalism. The concept was coined by philosopher Arne Naess in 1972 as an approach characterized by an intensely felt connection and a sense of solidarity with the non-human natural world. 'Biophilia' is a term popularized by Edward O. Wilson that describes a 'love of life or living systems' and the strong drive to affiliate with non-human nature. For deep ecologists, the experience of biophilia provides the moral energy necessary to do the challenging work of confronting ecocidal development. Deep ecologists typically consider a spiritual orientation advocated by Naess as a foundation for their work. Naess and other deep ecologists advocate direct action as a means to confront industrial development threatening ecological destruction.

Deep ecology has been criticized for its apparent essentialist perspectives; for putting too much emphasize on self and not enough on social forces; its biocentrism, its holism and its political naivety. Murray Bookchin offered this scathing critique:

> let me stress that deep ecology's limited, and sometimes distorted, social understanding explains why no other 'radical' ecology philosophy could be more congenial to the ruling elites of our time. Here is a perspective on the ecological crisis that blames our 'values' without going to the social sources of these values. It denounces population growth without explaining why the poor and oppressed proliferate in such huge numbers or what social changes could humanely stabilize the human population. It blames technology without asking who develops it and for what purposes. It denounces consumers without dealing with the grow-or-die economy that uses its vast media apparatus to get them to consume as a monstrous substitute for a culturally and spiritually meaningful life. (1993, 129–130)

In terms proposed in this book, deep ecology is like ecological literacy Mode One (see pp. 81–82) as it advances an experiential mode of engaging with ecological thought and practice: an approach that often fails to engage critically and politically. Nonetheless, the movement's dedication to a new understanding of human–nature relations (a relational ontology) and its explorations of the psychological dimension of emerging ecological awareness have been valuable contributions to ecological debates, theory and sensibilities. Its tenants appeal to many seeking alternatives to mainstream ecological modernization discourses with an approach that feels more meaningful and its followers often show immense courage using direct actions to protect forests, the climate, whales, and so on.

Permaculture + The Transition Movement

Permaculture is based on creating localized systems reliant on renewable energy and 'permanent agriculture'. It is a community design philosophy based on ecological principles and ethics. The term 'energy descent', coined by Howard and Elizabeth Odum and permaculture co-founder David Holmgren, refers to a decline in net energy supply. The flow of conventional fossil fuels is already declining in most regions due to the increasing scarcity of easy to access reserves.[5] Unconventional fossil fuels are now being extracted (fracking and tar sands) with more severe ecological consequences than conventional fuels and much higher costs. The concept of a managed process of energy descent is a radical departure from mainstream environmental thinking on energy which typically looks to decarbonize but not to dramatically reduce use. Energy descent is a central idea in permaculture and the Transition movement.

The Transition movement started in South West England in 2005. It draws on permaculture and deep ecology and works towards community design strategies in response to peak oil and climate change. Transition initiatives develop local 'energy descent action plans' as timetabled strategies for weaning a locality off fossil fuels and increasing local resilience in the face of change. The movement excels at community engagement and social learning with a variety of practices to make the difficult work of responding to environmental problems personally fulfilling for community activists. As transitioners work to re-localize by build regional resilience, they encounter the structural obstacles to sustainability. As these obstacles are confronted, the need to address the larger development model becomes obvious. With this process, the movement can also function as a politicization process. The transition movement is active with hundreds of local official initiatives worldwide.

Anarchism + Social Ecology

Probably the most misunderstood (and most often maligned) movement is that of anarchism. Noam Chomsky describes the anarchist tradition as 'a highly organized society, nothing to do with chaos, but based on democracy all the way through. That means democratic control of the workplaces, of federal structures, built on systems of voluntary association' (2013a, 107). It is a tradition that expects any hierarchical structure to justify itself. Otherwise it is 'illegimate and should be dismantled' (Ibid., 33). Anthropologist and anarchist activist David Graeber describes anarchism as

> a movement about reinventing democracy. It is not opposed to organization. It is about creating new forms of organization. It is not lacking in ideology. Those new forms of organization are its ideology. It is about creating and enacting horizontal networks instead of top-down structures like states, parties or corporations; networks based on principles of decentralized, non-hierarchical consensus democracy. Ultimately, it aspires to be much more than that, because ultimately it aspires to reinvent daily life as whole. (2002, 70)

The anarchist critique of authoritarian forms of government and elite power is supported by extensive anthropological work that explains how wealth is generated historically. Graeber's *Debt: The First 5,000 Years* (2011) describes the ways that debt has been historically constructed to serve the interests of elites. This timely work illustrates how austerity is a constructed scarcity: it is a consequence of specific economic policies – some of which resulted in the banking crisis and bailout of 2008 and subsequent further polarization of wealth.

Peter Kropotkin (1842–1921) was a geographer, an ecologist and an anarchist theorist who rejected Thomas Malthus' (1766–1834) theories on overpopulation and Charles Darwin's (1809–1982) theories on competition and survival of the fittest. In his book *Mutual Aid* (1903) he challenges the centrality of conflict in interpretations of evolutionary biology.

> Kropotkin famously contested Social Darwinist misappropriations of evolutionary theory, particularly the buttressing provided to laissez-faire capitalism by the ideology of 'the survival of the fittest.' Kropotkin acknowledged that there is competition and struggle in nature but, following Darwin, distinguished between (i) competition between organisms of the same species for limited resources and (ii) organisms struggling against the environment, which often leads to cooperation … *Mutual Aid* stressed the conventional underemphasis in Darwin's writings – and, more so, in conventional readings of Darwin – on the co-operative elements in evolution. The book then follows up with the suggestion that legacies of altruism and co-operation represent a clear thread running through social history in the form of free associations, communes, and cities. (White, Rudy & Gareau 2015, 27)

Kropotkin and other anarchist theorists hold that cooperative behaviour represents a force at least as powerful as competition in both non-human nature and in society. This argument is supported by the findings of modern biology, two of the most prominent writers on this theme are physician Lewis Thomas and biologist EO Wilson.

Critics are often left baffled by the anarchist vision of a society without a state. Anarchism has been criticized for offering no coherent vision of how we would keep the lights on and feed seven billion people as we build capacities for new decentralized democracy. It is harder to dismiss the anarchist critique of power and corruption in centralized and increasingly authoritarian governments. The anarchists' vision of regional eco-harmonious cities wherein the community or the city, not the state, is the focus for political activity, is based on the claim that centrally imposed organization does not work to promote harmonious relations or to effectively manage society. Anarchist theorist Graham Purchase claims that centralized power is 'incapable of preserving the regional ecological integrity so vital for planetary survival' (1997, 108). For anarchists, self-organization is the fundamental principle. They seek to build new ways of organizing social relations from the bottom upwards. This is performed by creating commons with social structures that are co-created and co-owned by collectives who share a commitment to making them work.

Anarchists see the state and capitalism as inherently dysfunctional and oppressive. Capitalism encourages individual greed and the exploitation of both human and ecological resources and is 'ill-suited for ecological needs of the planet' (Purchase 1997, 14). In working towards the construction of social and political life along a regional basis, anarchist theory suggests that

> Our industrial practices must follow a new non-capitalist, non consumerist and more bio-cooperative path in which the primary economic, political and social question is not: 'how many dollars will this make my city or country?' but 'how can we provide for humankind's most basic industrial needs while positively improving the ecological health of the nation in which industry is located?' (Ibid., 15)

Despite sensible nature of these insights, the anti-state and anti-capitalist anarchist critiques are typically dismissed or ridiculed in mainstream environmental discourses where anarchists are constantly misrepresented and maligned. Nevertheless, historically anarchists have made significant contribution to cooperative and labour movements and struggles for environmental and social justice. Anarchism practices have enabled more robust forms of democratic decision-making and organizing practices. Social movements have benefited from practical anarchist experience in horizontal decision-making processes and mutual aid networks. Anarchists have supported theoretical and practical efforts with a regional focus on green cities, alternative energy and small-scale agriculture for more than two centuries (Ibid., 73). Anarchists have historically been on the front line of anti-fascist organizing and resistance.

Social ecology is a theory initiated by Murray Bookchin who self-identified as anarchist for most of his life. Like anarchism, social ecology sees the environmental crisis as a result of social hierarchies and modes of domination. It rejects environmental discourses emphasizing scarcity and overpopulation with their implication that environmental problems are natural phenomenon. Instead Bookchin argues that environmental problems emerge from our politics. Changes in system structures, social relations and cultural attitudes can be a basis for an ecological society with enough for all. Directly democratic institutions and new liberatory eco-technologies will enable this transformation.

Eco-Marxism, Eco-Socialism + Communism

Eco-Marxism sees the environmental crisis as emerging from the ways that growth is achieved in capitalism. Like Kropotkin, Karl Marx (1818–1883) rejected Malthus' population doctrine. While much of the Marxist tradition has been ambiguous on the environment and certainly most historical and existing socialist states have records that are as bad or worse than capitalist states,[6] theorists such as Barry Commoner, David Harvey, John Bellamy Foster and Ted Benton have focused on the ecological aspects of Marx's work and further developed eco-socialist theory and practice. John Bellamy Foster describes a metabolic rift between capitalism and the Earth. The economy functions as 'a treadmill of production' wherein quantitative economic growth is structurally dependent on increases in production, resource consumption, energy use and pollution (Foster 2002). Growth is designed into the capitalist system. Profit-seeking firms exploit ecological and human resources by extracting value for the lowest cost possible. While profits are privatized, the harms done to the environment impact society as a whole (they are socialized). Any barriers to growth are slowly eroded as capitalism expands into domains that were once beyond market forces. Economic growth in capitalism depends on resource extraction and pollution – but the Earth's capacity to supply natural resources and absorb pollution is limited. It is the type of growth, not growth itself, which the problem. For example, Commoner documented a 53,000 per cent increase in nonreturnable soda bottles in the United States between 1946 and 1970 (1971, 143). If this growth in bottles had been qualitatively different, say in reusable glass, the environmental impact would have been dramatically less. During the same time period, returnable beer bottles production fell 36 per cent (Ibid., 143), so clearly it is the *types* of goods that are produced to meet human needs and desires that cause environmental harms.

Ecosocialists consider mainstream environmentalism's focus on consumption habits as largely ineffective. The decisions that matter (those with the most severe ecological and social impacts) are made by industrialists and policymakers. For example, in the context of air travel:

> At the broader level, an individual decision to fly less will have modest to little effect on greenhouse gas emissions if the airplane still takes off. We can, as individuals, decide to consume less but if the vast majority of waste comes from the productive side of the economy or if national economic policy is centrally dependent on more production and more consumption to 'get the economy going again,' this may become self-defeating. (White, Rudy & Gareau 2015, 92)

This focus on profit-seeking system structures rather than individual behaviour or shopping habits is a contribution that has been influential well beyond those who identify with socialism. Mainstream economists who have taken time to observe and analyse the way our economic system impacts the environment have often had to acknowledge the validity of this critique. Even establishment economist Lord Stern famously described climate change as 'the greatest and widest-ranging market failure ever seen' (2007, p. i). With the *Stern Report* (2007) the links between climate change and 'market failure' (due to the capitalist structures and dynamics) entered mainstream discourse on the environment.

The ecosocialist proposal is that deeper transformations in the political economy will need to occur to address both environmental problems and social problems created by the internal contradictions of capitalism. Most radical movements and a growing number of mainstream organizations now agree with this basic assertion. One valid critique of ecosocialism by the mainstream commentators (and also by the anarchist tradition) is that large governments associated with socialism always tend towards authoritarianism. Anarchist theorists further argue that the state is inherently capitalist and thus every state, even a socialist state with a strong regulatory regime, must function according to market logic in some way. In response to this problem, many contemporary communists share a vision of an anarchist version of Marxism where power is devolved with democratic regional decision-making bodies substituting for a traditional state.

Agroecology and Food Sovereignty

Agroecology is both the science of sustainable farming and a political movement concerned with food production, processing and distribution. Agroecology and food sovereignty aim to give small-scale farmers greater control of land, seeds, markets and labour. This is vital work because small-scale and family farmers feed more people globally than corporate agriculture (FAO 2014a). Family farms make up 98 per cent of farming holdings (FAO 2014b, 3) and supply 80 per cent of the world's food supply (FAO 2014b, 12). The traditions within agroecological food production are an antidote to some of the most serious environmental problems caused by unsustainable farming methods. For example, these methods rely largely on renewable resources available on the farm such as natural predators for pest control rather than external inputs like chemical fertilizers and toxic herbicides (such as neonicotinoid pesticide associated with bee colony collapse disorder). Agroecological food systems build resilience to climate destabilization and promote self-sufficiency in local food systems. Food sovereignty is a response to food security policies favouring corporate agriculture over small-scale and/or traditional producers. The result of food-security-oriented policy, according to food sovereignty campaigners has been dispossession (land grabbing in the Global South) and ecological degradation due to unsustainable industrial farming methods. La Via Campesina are a globally linked network of over 200 million peasants and small and middle-scale farmers, agricultural workers, landless people, rural women and indigenous communities in Asia, Africa, the Americas and Europe. They developed the concept of 'food sovereignty' to defend farmers' rights on issues of seeds and agrarian reform, resist corporate food regimes and industrial agriculture – while they simultaneously work to create sustainable food systems. The movement maintains that those who produce, distribute and consume food should control food system mechanisms and policies. La Via Campesina campaigns widely and maintains that small-scale traditional farming can feed the world (without corporate agriculture) in ecologically restorative ways that can also help bioregions adapt to climate change.

Indigeneity

By their own self-description, the natural environment is central to indigenous[7] culture and identity. Tradition indigenous worldviews see the Earth as living and animate along with everything in it as alive and filled with spirit. Soul or spirit is central to existence and this informs indigenous relationships with the land. A passage in the UNEP (United Nations Environment Program) complementary contribution to the Global Biodiversity Assessment describes indigenous perspectives:

> Although conservation and management practices are highly pragmatic, indigenous and traditional peoples generally view this knowledge as emanating from a spiritual base. All creation is sacred, and the sacred and secular are inseparable. Spirituality is the highest form of consciousness, and spiritual consciousness is the highest form of awareness. In this sense, a dimension of traditional knowledge is not local knowledge, but knowledge of the universal as expressed in the local. In indigenous and local cultures, experts exist who are peculiarly aware of Nature's organizing principles. (Posey 1999, 4)

The resulting commitment to particular geographic spaces has placed indigenous peoples at the forefront of movements defending land around the world. In its short history in Canada the Idle no More movement has resisted pipelines, tar sands and other extractive developments in Native territory. In the United States, Ojibwe activist Winona LaDuke describes the indigenous dedication to land: 'We know that's our territory and there's no other place to go. People don't get that, this is it' (2014). The imperative to protect that which sustains life is evident for many indigenous people and this is the focus of the campaigning and activism of the Indigenous Environmental Network and other many indigenous organizations. The many ways in which indigenous peoples have articulated that a balanced, reciprocal relationship with one's habitat are all contributions to the sustainability agenda. In many cases, the intensity of the indigenous struggle for their land is informed by the lived experience of being part of cultures that have been subject to genocide. Today a significant number (115) of at least 908 documented murders of environmentalists (over the 2002–2013 period) (Global Witness 2014, 10) were indigenous.

Indigenous activists offer some of the most challenging critiques to development policy on issues of environmental justice. The World People's Conference on Climate Change and the Rights of Mother Earth in Cochabamba Bolivia 2010 was a response to what they understood as the failed United Nations Climate Conference at Copenhagen (COP15). It was attended by around 30,000 people from over 100 countries. Two years later indigenous activists organized at Rio+20 2012 where the Kari-Oca 2 Declaration was published. This document condemned the United Nation's green economy project as

> a perverse attempt by corporations, extractive industries and governments to cash in on Creation by privatizing, commodifying, and selling off the Sacred and all forms of life and the sky, including the air we breathe, the water we drink and all the genes, plants, traditional seeds, trees, animals, fish, biological and cultural diversity, ecosystems and traditional knowledge that make life on Earth possible and enjoyable. (para 6)

Indigenous groups have been at the forefront of this campaign against the UNEP's Green Economy. Their territories are often those most threatened by the emerging policy decisions on biodiverse regions (see Chapter 13). Indigenous campaigners almost consistently assert a perspective that prioritizes ecological well-being and describe their own traditions as offering alternatives:

> We continue to inhabit and maintain the last remaining sustainable ecosystems and biodiversity hotspots in the world. We can contribute substantially to sustainable development but we believe that a holistic ecosystem framework for sustainable development should be promoted. This includes the integration of the human-rights based approach, ecosystem approach and culturally sensitive and knowledge-based approaches. (Kari-Oca 2 Declaration 2012, para 26)

While there is controversy about the relative degree of long-term sustainability of various indigenous societies, what is evident is that many of these cultures have existed without the ecological devastation done by the Western civilization in 200 years. The indigenous critique of the Western tradition is thorough. Native American Russell Means asserts that it is not capitalism, but the entire European tradition, that needs to be resisted. For Means, this resistance is about survival: 'We resist not to overthrow a government or to take political power, but because it is natural to resist extermination, to survive. We don't want power over white institutions; we want white institutions to disappear' (Means 1980). Note that Means is explicit that this is 'not a racial proposition but a cultural proposition' (Ibid.) that asserts that white institutions, white privilege, colonialism, imperialism and the domination agenda need to be dismantled for the survival of Native American peoples and the natural spaces that they depend on.

Ecofeminism

Ecofeminism reveals the conceptual artifices that present knowledge as universal and value-free. It describes how the logic of domination has historically been used to oppress both women and nature. The intersectional women's movement provides examples of large-scale social change to reorient society around new values. The tradition of critical consciousness in the women's movement demonstrates how particular practices can be used to transform circumstances and social relations. Ecofeminism calls for the next social transformation to enlarge the boundaries of concern to include the natural world (see pp. 64–65).

Another World Is Possible

The movements described here all contribute to social change for more ecologically viable and just futures. Groups in these traditions use diverse strategies to facilitate transition, often towards post-capitalist ways of organizing social relations. Greens, deep ecologists, Transition Town activists, anarchists, eco-Marxists, communists, agroecologists, indigenous peoples, ecofeminists, race and decolonization activists and others often work together resisting new ecologically destructive infrastructure; privatizations and corporate enclosures; anti-democratic institutions and policies; secretive trade agreements; land-grabbing and evictions; climate change enabling policies and other social and environmental injustices. These movements have helped create cooperatives, campaigns, marches, organizing strategies, mutual aid networks, labour movements, independent media, teach-in, strikes, boycotts, occupations, direct actions, refugee kitchens, protest camps and more. They do this to challenge and transform unjust and anti-environmental laws and activities while simultaneously building alternatives. They have won many battles to advance social justice and environmental sustainability and become increasingly sophisticated in their analysis and practice. New tools and strategies accelerate their work. In the past decade Occupy succeeded in changing the public discourse in the wake of the last financial crisis on the legitimacy of austerity and coined the 99 per cent versus 1 per cent meme that captured a new era of class conflict. Despite ongoing challenges, feminist principles have permeated all social change movements (to various degrees). New generations of women have freedoms our mothers and grandmothers fought to make possible. Indigenous people have been on the front line resisting ecologically destructive industrial development in some of the remaining most biodiversity-rich spaces on the planet. They have also produced some of the most scathing and insightful critiques of colonialism and neoliberal development policies. The struggle continues in thousands of local spaces as these movements are all fighting uphill battles – especially in the wake of the growing visibility and power of authoritarian governments and fascist movements. Social movements are networks and sites of social change experimentation and even survival for many vulnerable people. Socially responsive design can learn from this values-based work – often situated in, or linked to, communities under threat.

Notes

1. Calculated by the ecological footprint metric.
2. Examples of social movements are: Transition Towns, Occupy, Climate Camp, EarthFirst!, Via Campesina, Idle No More, Rising Tide and far too many others to list – see Paul Hawken's book in this subject, *Blessed Unrest* (2007).
3. Examples of NGOS (Non-Governmental Organization) are: Greenpeace, Friends of the Earth, WWF, Sierra Club, 350.org and far too many others to list.

4. The dynamic that motivates the pursuit of profit and the amassing of capital.
5. This peak in production is known as 'peak oil'.
6. Cuba is an exception and has been widely used in the Transition movement as an example of a state that faced energy descent during in the 1990s when their oil imports from the USSR collapsed. This case study was documented in the film *The Power of Community: How Cuba Survived Peak Oil* (Faith Morgan 2006).
7. Preferred names vary by location: Native American (USA), First Peoples and Aboriginal peoples (Canada), Inuit, San, Mbuti, and so on. There are at least 7,000 indigenous societies around the world (Hughes 2011, 14).

9   Ecological Perception 1 – Theory

Communication designers have a pivotal role to play in the creation of sustainable futures due to their ability to support the development of new cognitive skills for dealing with complexity and new social capacities to act on the basis of new knowledge. Designers can make ecological processes and relationships visible, tangible and accessible. By drawing attention to relationships, they can nurture ecological ways seeing and knowing. Within the context of an increasingly visual culture, images can communicate complexity and encourage the noticing of ecological phenomenon. Graphic design is especially well suited to facilitate ecological learning as it can draw on a wide variety of visual strategies to display specific geographic spaces, ecological patterns and processes, and complex dynamic systems. Environmental communication scholar Julie Doyle maintains that photography records circumstances of the past, so its usefulness in communicating climate change is limited to displaying damages already done (2009). Graphic design, on the other hand, has greater potential to respond to environmental communication challenges due to its ability to visualize information in very specific ways (this includes visualizing abstract concepts, future scenarios, geographic spaces, comparative data sets, complex systems and more). Designers use maps, charts, diagrams, graphs, timelines, illustrations, network visualizations, data visualization and information graphics towards different ends. In the following two chapters, I describe how designers can support new ways of seeing by directing attention to relationships – and thereby encourage relational or ecological perception.[1] The first chapter introduces the theory and the second the practice of ecological perception. To develop this concept, I review the ways in which perceptual practices have evolved over time with a brief history of communication theory and the emergence of 'visual intelligence'. I describe how visual metaphors function to challenge and potentially disrupt preconceived assumptions. Linking communication theory with ecological theory informs new strategies for revealing ecological relations. By displaying relationships, patterns and dynamics in complex systems, designers can support visual intelligence, relational perceptual practices and the ability to 'see systems'.

History of Modes of Communication

Social theorists and historians describe changing modes of communication and media as having profound social consequences. Famously, Marshall McLuhan explained, 'the effects of technology do not occur at the level of opinions or concepts, but alter sense ratios or patterns of perception steadily and without any resistance' (2001 [1967], 290). The historical shift from oral to written cultures in the West (starting in Greece around the fifth century BC) shaped new ways of perceiving and thinking that emerged in this time period. Historian Walter Ong describes oral cultures as having very different perceptual practices than those that characterize written cultures: in oral culture knowledge is fixed, formulaic and mnemonic (1982, 24). Reading and writing, on the other hand, are dissecting processes that organize information in a linear sequence and establish a context-free mode of communication, thereby conditioning consciousness towards an alienation from context and place, Ong's analysis of the seismic shifts that occurred during the historical oral–written communication revolution informs an understanding of an equally dramatic contemporary communication revolution from the written to the pictorial, iconic and digital. The awareness that human perception changes due to changes in modes of communication has significant implications for environmental learning in an increasingly visual contemporary culture.

The acute visuality of contemporary culture has led many visual theorists to support the concept of a contemporary pictorial turn (Dondis 1973; Mitchell 1994; Barry 1997). Others extend this idea and propose that visual culture is contributing to the emergence of new cognitive capacities (Barry 1997; Horn 1998; Chabris & Kosslyn 2005). Some see dangers in a pictorial turn. Chris Hedges describes an image-based culture as no longer having 'the linguistic and intellectual tools to cope with complexity, to separate illusion from reality' (2009, 44). Visual scholar Ann Marie Barry notes the danger of visual culture if approached uncritically:

> Today advances in technology have given commercial and political interests the ability to manipulate the way we see and comprehend our world before our understanding of visual images has fully matured. Few of us today understand how perception works, how cognition and emotions function, and how manipulations of attitudes, ideas and values can be accomplished. Fewer still, it seems, appreciate the implications of this. (1997, 333)

While visual culture can be used to mislead, the positive potential of visual culture is the opposite of its dangers: the benefits lie in its capacity to reveal complexity, convey strong messages, assist memory and issue-awareness (Nicholson-Cole 2005, 258). As with all potentially powerful tools and methods, critical thinking is essential. Images are effective tools to convey meaningful messages and facilitate learning but they can also be used to promote anti-ecological ways of thinking and conspicuous consumption. Visual literacy has been a priority for critical educators ever since Donis Dondis first articulated the need for a critical reading of images in *A Primer of Visual Literacy* (1973). Visual literacy is essential for both image-makers and the wider public who are bombarded with visual messages on television, online, outdoor advertising and in print.

Visual Intelligence

The concept of visual intelligence was developed by Ann Marie Barry, who describes it as necessary not only to resist the influence of messages absorbed uncritically, but also to develop the capacity to think in abstract and perceptually oriented ways (1997, 7). An awareness of the logic of visual messages is important because visuals are a powerful force of socialization. Visual intelligence extends the range of human communication and ways of knowing. Barry explains that perception is the basis of knowing and understanding – but our relationship to perception is naïve. Too often, she says, 'to see is to believe' (Ibid., 1). Visual intelligence involves a perpetual awareness that the visual world is an *interpretation* of reality, following Alfred Korzybski's first principle of general semantics; 'the map is not the territory' (1933, 750). Barry identifies the need for a critical function to interrupt and critically assess the perceptual process since perception 'is not a direct recording of what is out there, but a mental configuration that we interpret as an image – the end result of a highly exploratory and complex information-seeking system' (1997, 37). It is because visual representations are an interpretation of reality, and not reality itself, that images can also be used to challenge and re-interpret dysfunctional ways of knowing – such as those perspectives which perpetuate environmental problems.

Herein lies an opportunity for intervention with visual communication. When conceptual maps are acknowledged merely as representation, re-visioning more effective representations becomes possible. Visual intelligence as described by Barry is considered to be inherently relational and contextual. Images are capable of displaying non-linear, dynamical systems and revealing pattern (Ibid., 9) and are useful at displaying relationships. Visual intelligence lends itself to the types of competencies that support ecological literacy and ultimately ecological perception.

Design Intelligence

Visual intelligence is similar in many ways to the concept of design intelligence as described by Nigel Cross (1990) and others. Barry's and Cross' work both draw on Howard Gardner's (1983) ideas on multiple intelligences (Barry 1997; Cross 2010). According to Gardner, visuals are linked to spatial intelligence (skilled at mental imaging) and patterning intelligence (the ability to see patterns) (Barry 1997, 8). Visual intelligence enables greater understanding of patterns and dynamic systems. (This noticing pattern of relationship was advocated earlier by Gregory Bateson). Barry distinguishes between the logical and linear structure of verbal language and the non-linear and open structure of visual language and suggests that visual intelligence is a means of integrating these modes. Cross describes designerly ways of knowing as 'sensitive to nuance' (2010, 2), skilled at recognizing pattern, and offering 'a way of thinking that involves operating seamlessly across different levels of detail, from high level systemic goals to low level physical principles' (Ibid., 3). Design intelligence is an 'ability to shift easily and rapidly between concrete representations and abstract thought, between doing and thinking' (Ibid., 3). Like visual intelligence, design intelligence can be trained and developed. Both visual and design intelligence describe expanded, multidimensional perceptual capacities. Both support the types of competencies needed for ecological literacy.

Visual Language

The popularity of data visualization, network visualization, system mapping and other new visual tools suggest that visual culture is already supporting new perceptual practices. In *Visual Language* (1998), Robert Horn claims that the proliferation of visual communication indicates that we are witnessing the emergence of a new visual language that integrates words, shapes and images. This visual language is capable of expressing complex ideas more efficiently than anything we have yet known. Horn describes visual language as an antidote to the fragmentation and reductionism of current ways of communicating and as a means to deal with increasing complexity: 'Visual language has the potential for increasing human "bandwidth", the capacity to take in, comprehend, and more efficiently synthesize large amounts of new information. It has this capacity on the individual, group, and organisational levels' (2001, 1). According to Horn, human cognitive effectiveness is constrained by the 'limitations of working memory that George Miller identified in 1957' (1998, 4) but the integration of words, images and shapes within visual language enables new capacities to deal with increasing complexity, directly addressing these limitations of human cognitive abilities. Horn explains: 'without aids our minds are quite unable to cope with the sheer size, interrelatedness and complexity of projects' (1998, 45). Research suggests that images can expand human cognitive capacities to deal with multifaceted information, enabling synthesis of disparate information (Chabris & Kosslyn 2005, 36). Studies indicate that the more visual communication becomes, the more likely it is to be recalled: this is known as the pictorial superiority effect. Visual language is a means of negotiating increasing complexity by making it comprehensible.

Relational or Ecological Perception

The fact that perception is influenced by communication and that perceptual practices can be intentionally developed, is good news for sustainability communication and education. Bateson (1972) and Barry (1997) both describe perception as a learnt skill, which is 'fixed into patterns' (Barry 1997, 50). If this is true, visual communication can be designed with the explicit intention of supporting the development of new perceptual skills. Visual theorist Laura Sewall explains that it is the focus of attention that 'often means the difference between seeing and not seeing' (1995, 204). Perceptual attention and perceptual habits are a basis for our way of seeing (Sewall 1999). Through practice 'visual systems structures, or neural networks, determine our perceptual tendencies' (Ibid., 207). By encouraging new perceptual practices, new ways of seeing become possible. Designers can influence ways of seeing through directed selective attention, by visualizing phenomenon in ways that emphases relationships. This supports the noticing of relations and engages relational, or ecological perception. In this way, design can support ecological literacy through emphasizing 'the patterns that connect' (Bateson 1972). The understanding that perception is highly dependent on what we already think we know (or what we think we are looking at/for) enables change-making strategies.

Ecological literacy is developed through a process of dismantling erroneous ways of thinking based on the illusion of ontological atomization independent of ecological context. Epistemological error is reinforced through perceptual habits, which are in turn embedded within and further reinforced by social practices, media and communication. With the knowledge that perceptual habits can be changed, visual communication can be constructed to encourage relational perceptual habits and a 'shift in our way of seeing ... to affirm the complexities and mutual integration' (Bateson & Bateson 1988, 176). Ecological perception is developed through establishing these new perceptual habits that ultimately transform ways of seeing.

Ecological theorists Laura Sewall and David Abrams see perception as holding unique potential to address the denial of ecological relations. They characterize the environmental crisis as arising from a collective perceptual disorder (a kind of myopia) disabling the recognition of both ecological context and interconnectedness. Sewall's work on perceptual psychology and Abrams work on cognition and perception explores how the manner in which reality is seen or perceived affects understanding. Both theorists describe perceiving ecological embeddedness as central to ecological learning. The goal sounds simple enough: to encourage the perception that 'we are within the biosphere as opposed to on the planet' (Sewall 1995, 212). In practice, developing this awareness is easier said than done. The notion of depth is important here because the deeper the qualitative aesthetic experience, the deeper the potential associated shift in perception. Sewall and Abrams both use the depth metaphor as a means of describing a shift from perceptual blindness to an embedded perspective. Depth refers to the *perception* of ecological relations rather than an *intellectual* recognition of this idea as a theoretical construct. The deeper the perceptual experience, 'the greater the potential for transcending perceptual habit' (Ibid., 213). Perception evokes deeper modes of knowing that intellectual propositions and thereby enables more profound changes in worldviews, values and associated behaviours.

Ecological perception involves a shift from seeing *objects* to perceiving *interfaces between* objects. Ecological thought emphasizes context but current perceptual practices tend to focus on objects as atomized, isolated elements; 'we readily perceive things but are relatively insensitive to the relationships between them' (Ibid., 207). Sewall explains 'our attention focus … creates subjective reality by facilitating the perception of some objects, relations, and events to the exclusion of others' (Ibid., 207). Specific strategies that support ecological perception described by Sewall are:

Sewall's Strategies for Ecological Perception

1 learning to attend within the visual domain;
2 learning to perceive relationships;
3 developing perceptual flexibility;
4 learning to re-perceive depth;
5 the intentional use of imagination. (1995, 204)

These practices support a shift in focus of attention. Sewall's strategies support a shift in 'subjective reality by facilitating the perception of some objects, relations and events to the exclusion of others' (Ibid., 207). New perceptual practices can potentially 'extend our narrow experience of self … [and support an] identification with the external world' (Ibid., 204). Perceptive shifts occur 'when objects are seen in relationship, rather than within a kind of perceptual isolation' (Ibid., 209). Relational perception broadens and deepens understanding of ecological self.

The shift in perception sought by ecological theorists is one that reflects scientific findings. Over a century ago quantum physics demonstrated that on a quantum level the act of observation changes the phenomenon being observed such that matter meets the observer's expectations. The implications of this discovery are still in the process of being incorporated into visual theory and design practice. Our relationship with the world is mediated by the act of perception and the conceptual architecture (worldviews, paradigms, ideologies) that filter what we chose to notice. Ideally we attempt to see as clearly as possible what is actually before us with as little bias as possible, but it is wrong to assume that any of us is able to perceive the world without mental maps that help us make sense of what we are seeing. We are now in a situation where we need to build more effective mental maps, based on relational and participative ways of seeing, knowing and acting. This shift in perception can be facilitated by qualitative aesthetic experiences. Communication design, as a practice that facilitates ways of seeing, can help.

Visual Metaphor

Metaphors are one of the ways in which meaning is most powerfully constructed. Visual metaphors are considered to have more emotional impact than linguistic metaphors (Ortony 1979, 11) while multimodal metaphors allow for even subtler ways of communicating, 'hence achieving rhetorical effects more unobtrusively' (Forceville 2008, 477). Metaphors are used to communicate new ideas and establish cultural legitimacy by harnessing preconceived ideas to create new understanding. The human ability to understand new information is enhanced as new visual metaphors are developed. New visual metaphors make new information and ideas accessible and emotionally meaningful.

Visual metaphors work subtly to communicate new ideas in ways that are specific to a cultural context. Decoding visual communication is a learning process that involves deciphering the meaning of visual metaphors and their cultural connotations. We decipher visual communication on the basis of previously established visual codes. Psychologist and visual theorist Jan B. Deregowski's research with previously uncontacted people demonstrated that the tribe he worked with did not recognize a drawing of an antelope despite the animal being an everyday feature of their lives (1980, 113). Each culture has unique conventions and visual codes for representing the world. The ability to interpret visual communication, like oral and written language, is a culturally specific learning process. Visual communication shapes experiences and makes meaning as a result of connections between codes, metaphor and content that are culturally specific. Graphic designers organize analysis through culturally relevant visual metaphors in order to convey information and also create emotional impact. Metaphors are culturally constructed tools that convey meaning based on ideas that are already significant to audiences in ways that evoke sensibilities and feelings.

Metaphors are understood through cognitive frames that are the lenses through which people see the world. These cognitive frames are the result of our experiences, beliefs and assumptions. We are often unaware of our own cognitive frames and consider our perception to be simply how things are. This becomes a problem when basic premises are based on incorrect assumptions. Normally conceptual frames are taken for granted, so we don't examine them until they are recognized as problematic. The ecological crisis demands that we spell out cognitive structures as a means of resolving deep-rooted epistemological error. Deep frames exist within cognitive structures held in long-term memory. The theory of epistemological error suggests that deep frames do not reflect ontological embeddedness. The prescription for this dilemma is working explicitly with frames of reference.

Metaphors can help reframe our understanding of new information during times of transition. Historically, metaphors have dramatically shifted humankind's understanding of the natural world. The scientific revolution saw a complete change in ontology and epistemology from earlier periods replacing the geocentric view of the world with Bacon's empiricism, Descartes' rationalism and Newton's mechanism (see Chapter 4). Feminist historian Carolyn Merchant describes how the metaphors used by the fathers of the scientific revolution served to reconstruct human consciousness in the sixteenth to seventeenth centuries by reframing nature as a female, passive and as an object open for exploitation and domination. The dramatic changes in human–nature relations that occurred in this time period constituted 'the death of nature – the most far reaching effect of the scientific revolution' (Merchant 2001, 281). Ecological theorists (Bateson 1972; Shiva 1988; Capra 1997; Merchant 2001; Sterling 2001) describe the root metaphors on which the Western scientific establishment has been built as based on exploitation and violence towards nature.

Using metaphors in new contexts allows designers to establish new meanings. Visual metaphors can generate new frames of reference reflecting mutualistic relations with the non-human natural world. The feminist, indigenous and permaculture movements are leading this movement to develop generative (rather than exploitative) way of conceptualizing and relating to non-human nature. Since design is a practice that is often engaged with developing new ways of framing a problem, design can also help here. Within design 'framing is often seen as the key creative spirit that allow original solutions to be produced' (Paton & Dorst 2010, 317). Visual metaphors are a means to introduce new frames of reference. Designers are in a good position to introduce new visual metaphors supporting mutualistic relations with our ecological context.

Critical Visual Literacy

Working with frames must be done as transparently as possible to demonstrate how ideas are reproduced in images and other cultural products and processes. All communication design is embedded with assumptions. While the dominant ideology appears natural and is often not recognized as a point of view arising from historical, cultural and political circumstances – filtering mechanisms can be revealed. Mary Catherine Bateson explains that 'our patterns of attention and inattention are learned and can be changed … [but] if you want to change a pattern of thought, you may need to bring it into consciousness' (2005, 318). Within a cultural environment that has deeply embedded anti-ecological patterns of thought, thinking critically about communication practices is vital.

Visual communication has the potential to be harnessed to serve different priorities. All communication design fosters unique cognitive and perceptual habits. It can reinforce atomized, fragmented and isolating ways of perceiving the world. In this mode, design reproduces alienated ways of seeing and knowing that are de-contextualized and disconnected from material, historic and political realities. Alternatively, communication design can support the development of perceptual practices that emphasis context, participation and relationships. Barry emphases that it remains to be seen if visual media:

> and the profound emotional power they tap and release will be the province of individual intelligence and cumulative social wisdom, or it will remain exclusively in the realm of special interests groups driven by short-sighted economic or other motives. Visual intelligence ultimately implies an understanding of the power of the visual image both to reveal and to reshape the world. (1997, 337)

While images can be used to reveal ecological relations and engage ecological learning, more often they are often used to create elaborate marketing spectacles celebrating environmentally harmful activities.

Nurturing relational perception is ambitious role for designers. Ecological perception is a building block for sustainability because 'perception, consciousness, and behaviour are all radically interdependent' (Sewall 1995, 203). New perceptual habits can be facilitated by design – especially graphic, digital and interaction design in specific ways I describe in the next chapter. By encouraging the noticing of connections, interactions, influences and dynamics graphic design can function to dislodge conventional reductive approaches to knowledge. Breaking old patterns make space for new ways of seeing.

Note

1. I use relational and ecological perception interchangeably as both words have connotations I would like to evoke with this concept. I developed this concept in my PhD (Boehnert 2012) and subsequent papers (Boehnert 2014).

10 Ecological Perception 2 – Practice

Images have historically played an important role in facilitating learning (in science and elsewhere) by illustrating new ideas and phenomenon. Graphic designers make new information accessible through the selective framing and structuring of information to highlight relevant facts and ideas. Knowledge visualization captures as much relevant information as possible by creating overviews and displaying relationships while simultaneously displaying intricate details. Visualization practices are engaging with complexity in ways that move

> the focus from single parts to the whole, and sustains the need to understand the specific configuration of relationships (*pattern*). Understanding and intervening in a complex system requires the perception of the system as an integrated structure, and understanding which properties characterise the whole system, rather than belonging to any one specific component…those properties that emerge from the relationships and the interactions between single elements. (Valsecchi et al. 2010, 2)

The attention to relationships informs analysis of complex and dynamic systems and support relational and ecological perception. Designers are harnessing the latent potential of images to nurture relational and ecological perception by directing attention to context, causality, connections and comparisons. This relational focus reveals patterns. With attention to qualitative values and aesthetics, deeper perceptual experiences become accessible. These experiences can nurture the will to act on new knowledge. A relational, contextual and qualitative approach supports system thinking across different scales and sectors, the design of effective interventions and the desire to do this work.

Systems-Oriented Design

System-oriented design (SOD) is an expansive approach to design that supports learning on interconnected, messy problems that straddle environment-social-economic-political spheres. Since systems work is often facilitated through images, communication designers have valuable contributions to make. SOD focuses on relationships. Birger Sevaldson explains that with SOD the focus shifts 'attention from the objects and entities to the relations between them' (2013, 13). With more precise language on connectivity, SOD supports new ways of approaching of knowledge:

> The systems oriented designer is initially less concerned about hierarchies and boundaries of systems and more interested in looking at vast fields of relations and patterns of interactions. She is geared towards looking at as many interrelations as possible and working with a 'field-feel' and holistic overview, while making details accessible. The systems oriented designer is looking beyond the object (product or service) and she perceives the object merely as a 'symptom' or 'outcropping' of vast systems that lay behind the object. (Ibid., 3)

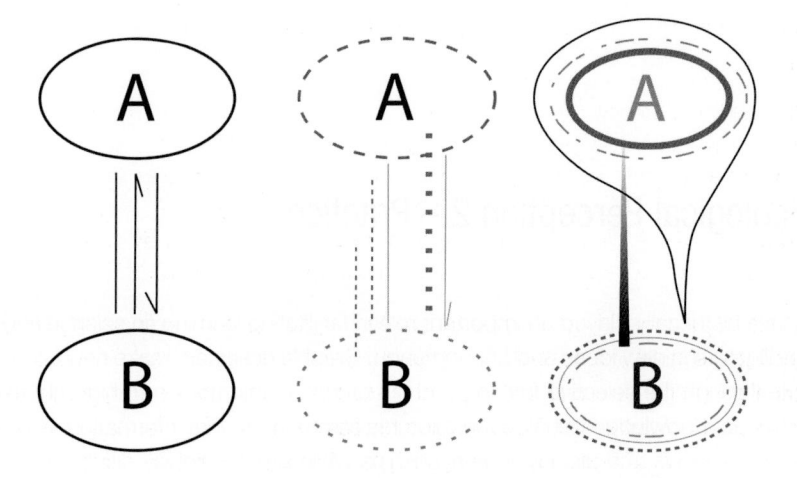

Fig. 10.1 Diagram of different ways to graphically represent relations between two entities. Line, colour and weight are used to codify the relationships (Birger Sevaldson, 2013).

Sevaldson has categorized types of relationships as structural, hierarchical, semantic, thematic, associative, representational, causal, flows, feedback loops (both negative and positive), spatial, temporal, rhythm, repetitions (Sevaldson 2016). Figure 10.1 displays some of the ways relationships can be coded visually (with line type, weight and colour). Describing the links between actors in a system or a network contributes to the theory of relationality. Sevaldson has also categorized types of systemic relations in Figure 10.2.

SOD typically starts with system mapping processes. Giga-Mapping is a system visualization practice for large information-dense diagrams that enable SOD.

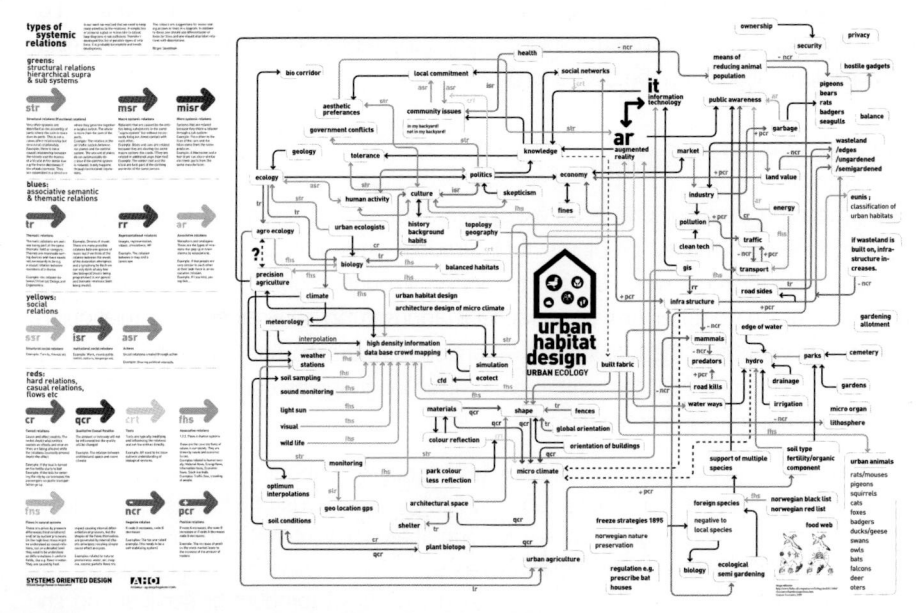

Fig. 10.2 *Types of systemic relations.* Birger Sevaldson, 2016.

122   Design, Ecology, Politics

Sevaldson describes how the mapping method contributes to analysis, synthesis and communication:

> The designerly approach to mapping made me aware of the importance of designing as a way of investigation and generation of visions that opened up the space for intervention. GIGA-maps therefore developed into design artefacts. This was an important step because this realisation connected designing and analyses and reasoning into one device. Designing was used in close relation to analysing and synthesising … By designing new questions emerged followed by new rounds of inquiries and mapping. (2013, 11–12)

Maps use visual strategies to capture as much meaningful information as possible by seeking richness over simplifications (Ibid., 5). Systems-oriented designers work to move beyond the descriptive to generate a new vision by interpreting and framing 'what is' and visualizing what 'what could be' (Ibid., 15). The mapping process is a means of exploring the design problem. For example, visually mapping a 'design problem including elements (people and resources), interconnections (relationships) and intended purpose (goal)' (Benson & Perullo 2017, 14). Mapping help designers expand the scope of their analysis on a specific design brief. Giga-mapping and other mapping techniques facilitate interdisciplinary collaborations and support self-adaptive learning for more effective systems design.

Addressing Epistemological Error by Design

With design, worldviews are illustrated. Working with visuals helps build an understanding of 'the power of worldviews to determine perception and ultimately reality' as it builds awareness of how images become powerful 'guides, or templates for the myriad unconscious decisions we make' (Sewall 1995, 214). As images work in this way, they can also be a means of challenging problematic frames of reference and facilitating new ways of thinking. Learning to perceive relationships (rather than isolated elements) is a new skill for us as we break the habit of perceiving things in perceptual isolation (Wheatley 2006. Traditional way of knowing narrow our field of awareness. Visual communication can facilitate expanded and relational perception by developing strategies that reveal:

1 **context** – reflecting embeddedness and focusing attention on spatial relations
2 **causality** – addressing unintended consequences, flows and remoteness
3 **connections** – illustrating interdependence and relationships between actors
4 **comparison** – displaying quantitative and/or qualitative information in relationships

The display of context, causality, connections and comparisons all support relational and ecological perception. In the following section I describe each strategy and provide examples.

Context

Context refers to both physical and conceptual space. Mapping practices illustrate context in ways that focus on specific details while also presenting the wider context. Maps display dynamics and individual elements that cannot be understood in isolation. By visualizing the whole, patterns become apparent (patterns are revealed with distance and perspective). It is this feature – of displaying the whole (sometimes revealing patterns) and the details (revealing individual elements) *simultaneously* that make maps so useful. Maps support capacities to think about whole systems and details at the same time: to hold attention at two (or more) levels concurrently. By moving back and forth between levels we can bring 'sensitivities and information gleaned from one level to help us understand the other' (Wheatley 2006, 143). Visualizations enable system wide overviews while also displaying individual elements or moments. For example, Greenpeace's oceans campaign by Density Design (Figure 10.3) displays multiple levels of information on the environmental impacts of fishing.

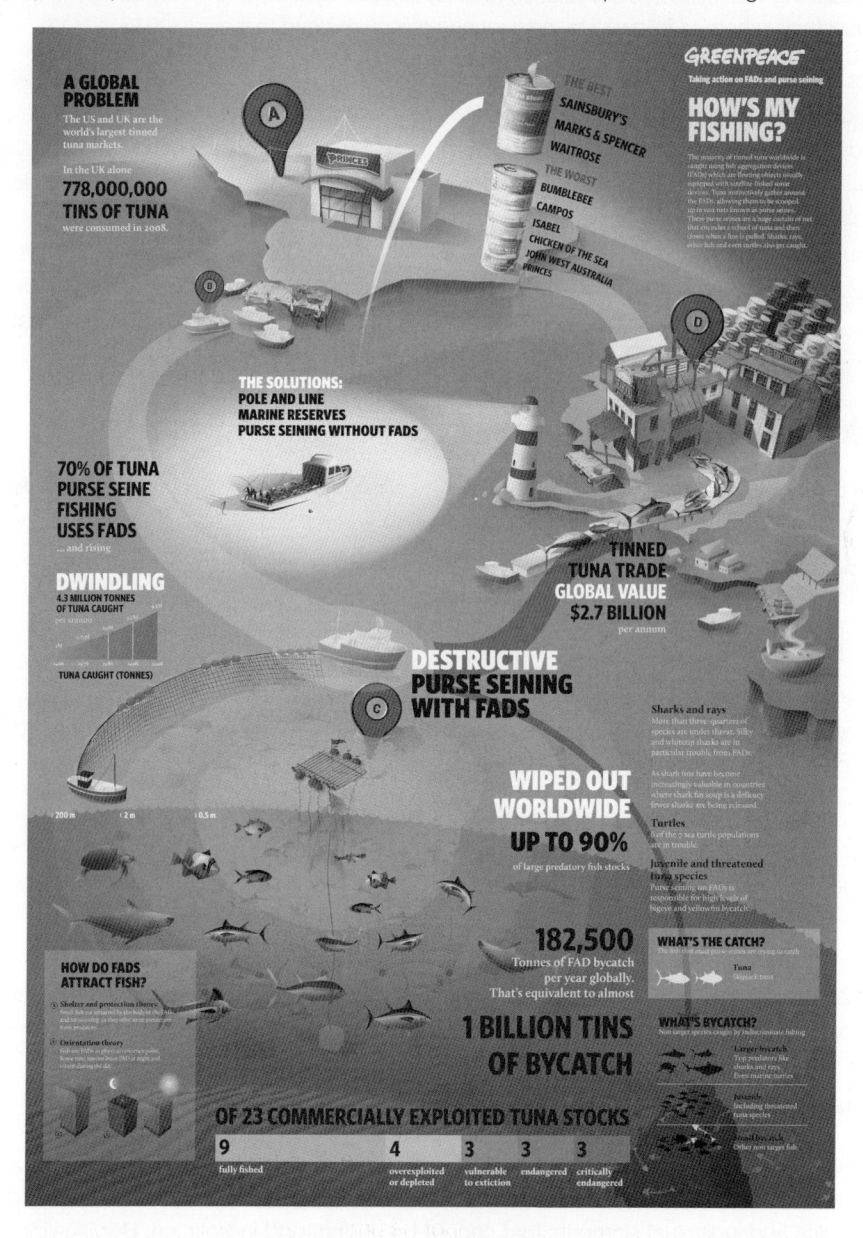

Fig. 10.3 *Greenpeace oceans campaign. How's my fishing?* Density Design and Greenpeace UK. Donato Ricci (Project leader), Michele Graffieti (main investigator), Luca Masud, Mario Porpora (designer); Greenpeace Team. Hannah Davey, Viola Sampson, 2010. Greenpeace UK.

GEOGRAPHIC MAPS

The tradition of geographic map-making offers a rich resource of visual strategies and devices to display context, relationships and hierarchy. Interactive and geographic information system (GIS) tools offer options for the visual communication of geographic content. Geographic maps support the development of situated knowledge and create meaning through displays of relationships between elements. With maps fine texture encourages 'micro/macro reading of detail and panorama' thereby potentially realizing both local and regional or global comparisons (Tufte 1990, 33). Geographic maps use colour, shade, line and icons as indication of depth, boundaries and types of geological or constructed features. High-density design allows readers to chose their own method for reading the map and focus on specific interests. With mapping strategies designers establish relationships and situate information in a particular location. Visualizing place focuses on local issues. The focus on specific place is important for ecological learning as ecological problems are context specific. Obviously context can also be global where mapping focuses on a particular global issue.

CONCEPTUAL MAPS

Maps can also be about ideas. Conceptual maps illustrate the structure of complex issues and communicate multidimensional information. Robert Horn lists a wide variety of types of conceptual maps: strategy maps, options maps, scenario maps, argumentation maps, cross-boundary causality and dynamics maps, unknown territory maps, public rhetoric maps, worldview influences maps, dilemmas map and paradoxes maps (2001, 4–5). These visualizations illustrate the 'logical structure and visual structure of the emerging arguments, empirical data, scenarios, trends and policy options... and help keep the big picture from being obscured by the details' (Ibid., 5). Conceptual maps can share mental models. They can be tools for communication emerging ways of thinking (consider for example how mind maps are used when brainstorming). When things are not functioning in productive ways, often the best thing we can do is to develop methods to learn about the dynamics of a situation. In the process of making a map, problems are sketched and analysed. Displaying the structure of complex issues builds knowledge, supports decision-making and enables more effective design interventions. With *The Small-Scale Energy Harvesting Project* map designer Francesco Zorzi illustrates ways of making small everyday objects into energy harvesting devices (rather than energy consuming items). This suggestion for the distribution of energy production uses a piggy bank metaphor to conceptualize new distributed energy systems (Figure 10.4).

Causality

Images can reveal causality by illustrating sequence of activities or chains of cause and effect. The visual vocabulary developed with system maps reveals dynamics and causality. Links are coded to define active processes that connect actors. Notations in network and causality diagrams enhance communicative capacities and supports systems thinking. Edward Tufte describes the importance of illustrating 'what cause provokes what effect, by what means, at what rate' (Tufte 1997, 9). Mapping causality can help direct attention towards unintended consequences and other unknowns. Since causality within the Earth sciences is complex, often the environmental consequences of specific actions cannot be

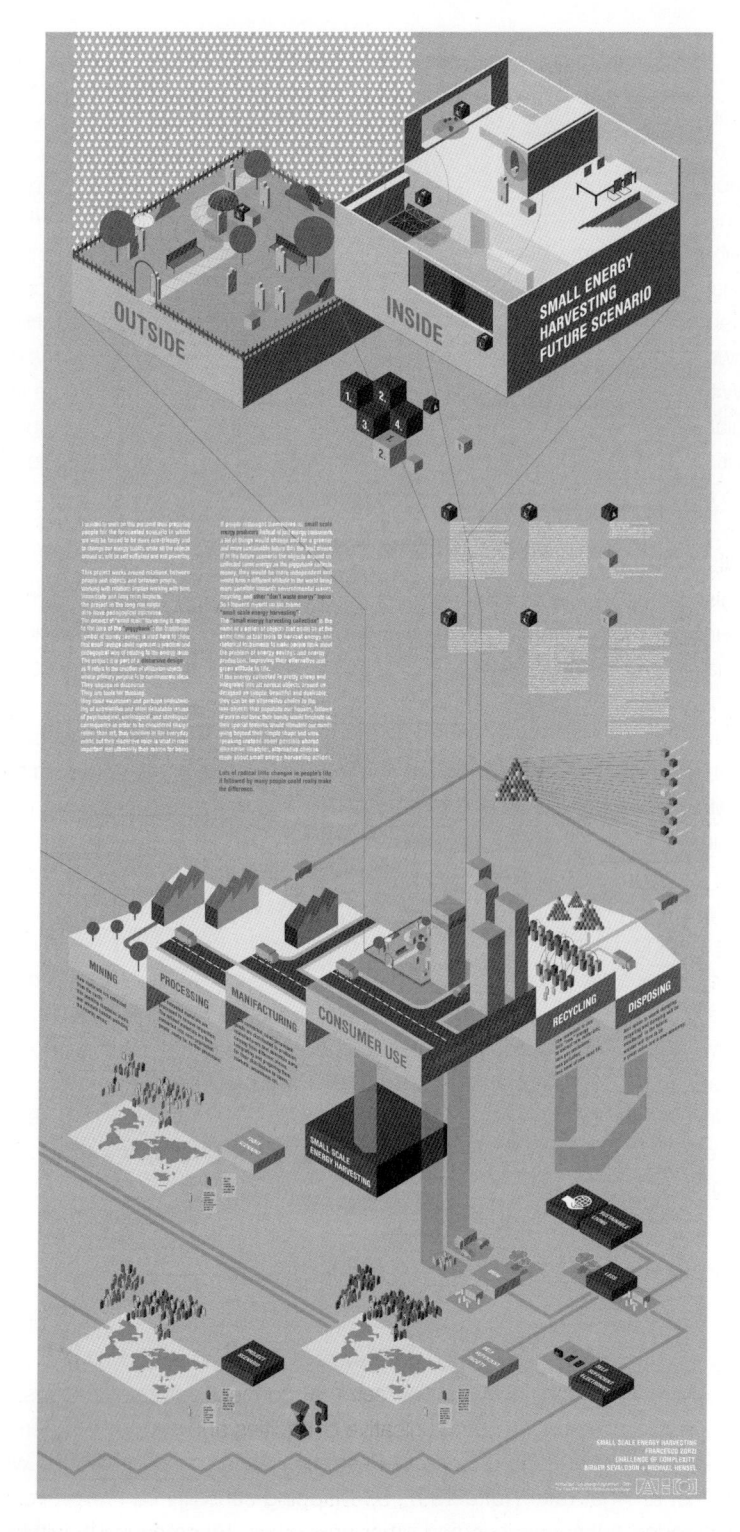

Fig. 10.4 The Small-Scale Energy Harvesting Project proposes making small everyday objects into energy harvesting devices. Artwork by Francesco Zorzi, tutor: Birger Sevaldson, 2009.

126 Design, Ecology, Politics

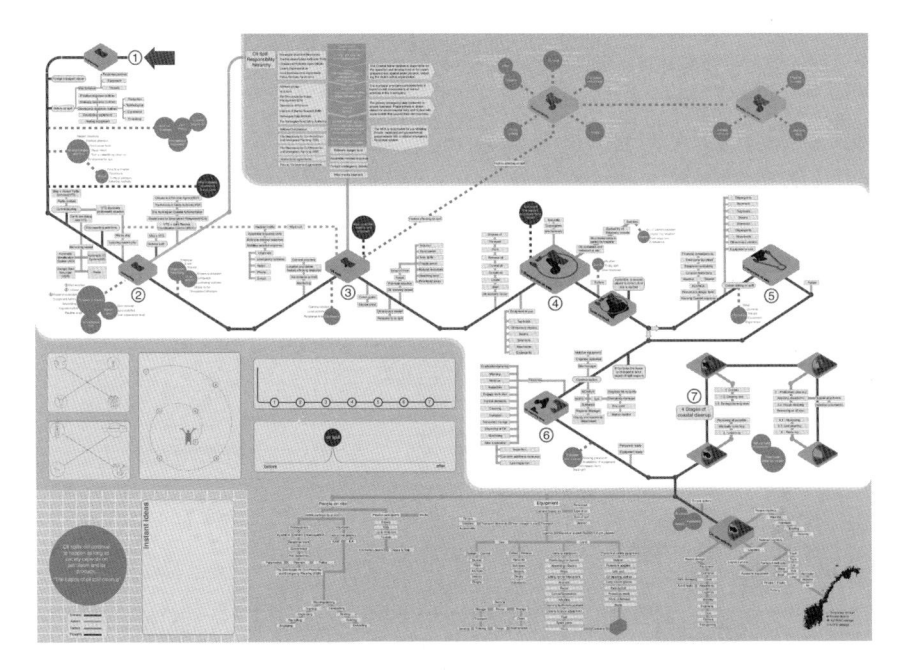

Fig. 10.5 Flow of actions during an oil spill accident when a ship runs on ground on the coast of Norway. The diagram displays actors and stakeholders involved and influenced by such an event. Moreover, it communicates risk and displays potential for new systems interventions to prevent accidents. (Master's Thesis by Adrian Paulsen, Advisor: Birger Sevaldson, Oslo School of Architecture and Design).

predicted. For this reason, images on issues of the environment should represent uncertainty in a manner that does justice to the potential dangers of unintended consequences. Design can create an appreciation for uncertainty, risk and ecological thresholds by drawing attention to these (often unknown) factors. Horn describes a role for the visual representations of unknowns by mapping gaps in knowledge as 'unknown territory maps'. Focusing on unknown information serves to illustrate 'the depth and breadth of our ignorance in this area so as to inform the debate about the precautionary principle' (Horn 2005, 2). Figure 10.5 displays a flow of actions during an oil spill accident off the coast of Norway. The diagram displays potential for new systems interventions. By making unknowns explicit, images can enhance an understanding of risk and can be a way of avoiding dangerous mistakes.

Connections

Network visualization displays webs of relationships. Unlike text they encourage undirected and interactive reading. Network visualization can display the nature of relationships between nodes (representing actors) and the connecting links and arches (representing interactions, structure and typology of the network), thereby revealing information that might be unavailable otherwise. Narrative can be condensed into a series of signs that indicate the nature of relationships between nodes. These visual codes make it possible to display dozens or even thousands of nodes and links. Manuel Lima's typology of network visualizations (Figure 10.6) illustrates various ways that networks are organized. Lima's website and book (*Visual*

Complexity: Mapping Patterns of Information) collect natural and human-made networks in biology, physics, ecology, computer science, sociology, news, government data and other fields. His work describes typologies, principles and concepts in the visualization of networks. Network visualizations such as these support relational perception by focusing on connectivity.

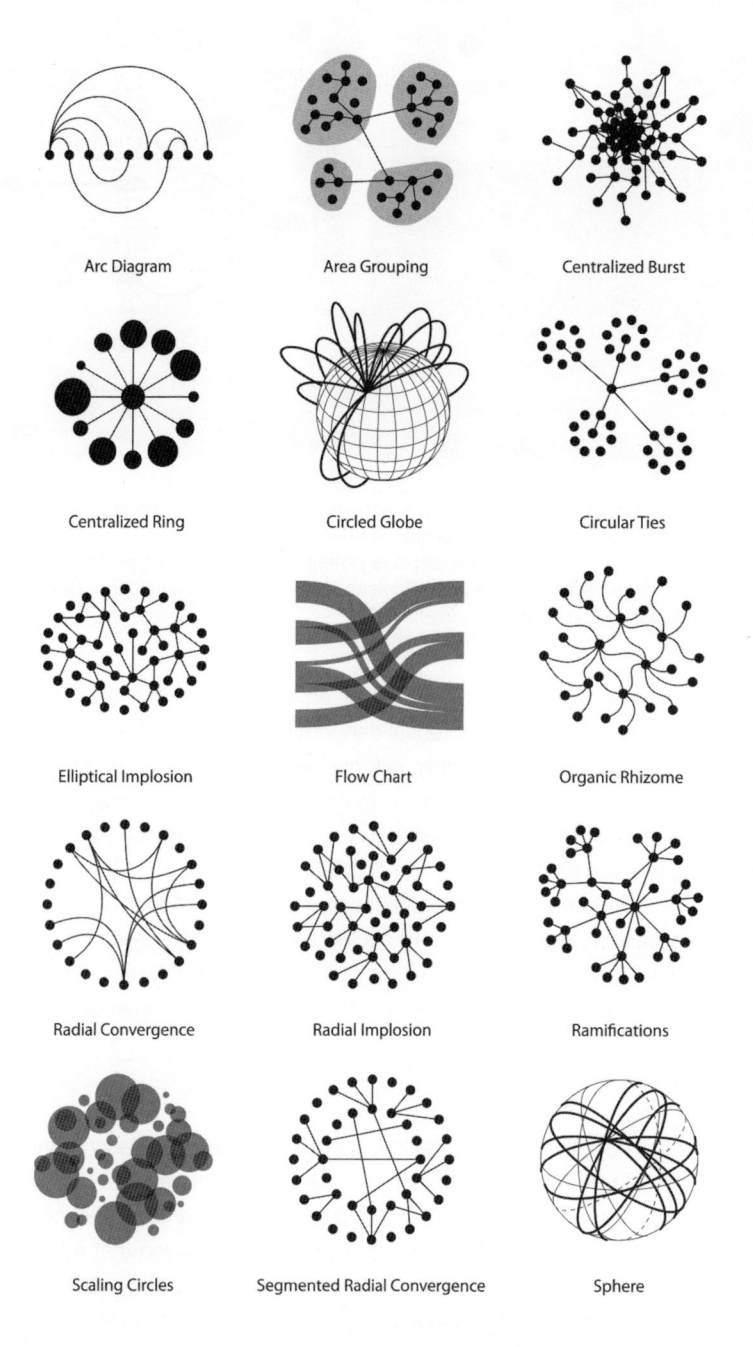

Arc Diagram   Area Grouping   Centralized Burst

Centralized Ring   Circled Globe   Circular Ties

Elliptical Implosion   Flow Chart   Organic Rhizome

Radial Convergence   Radial Implosion   Ramifications

Scaling Circles   Segmented Radial Convergence   Sphere

Fig. 10.6 *The Syntax of a New Language. Visual Complexity: Mapping Patterns of Information*. Manual Lima. Image displaying network structure, 2011.

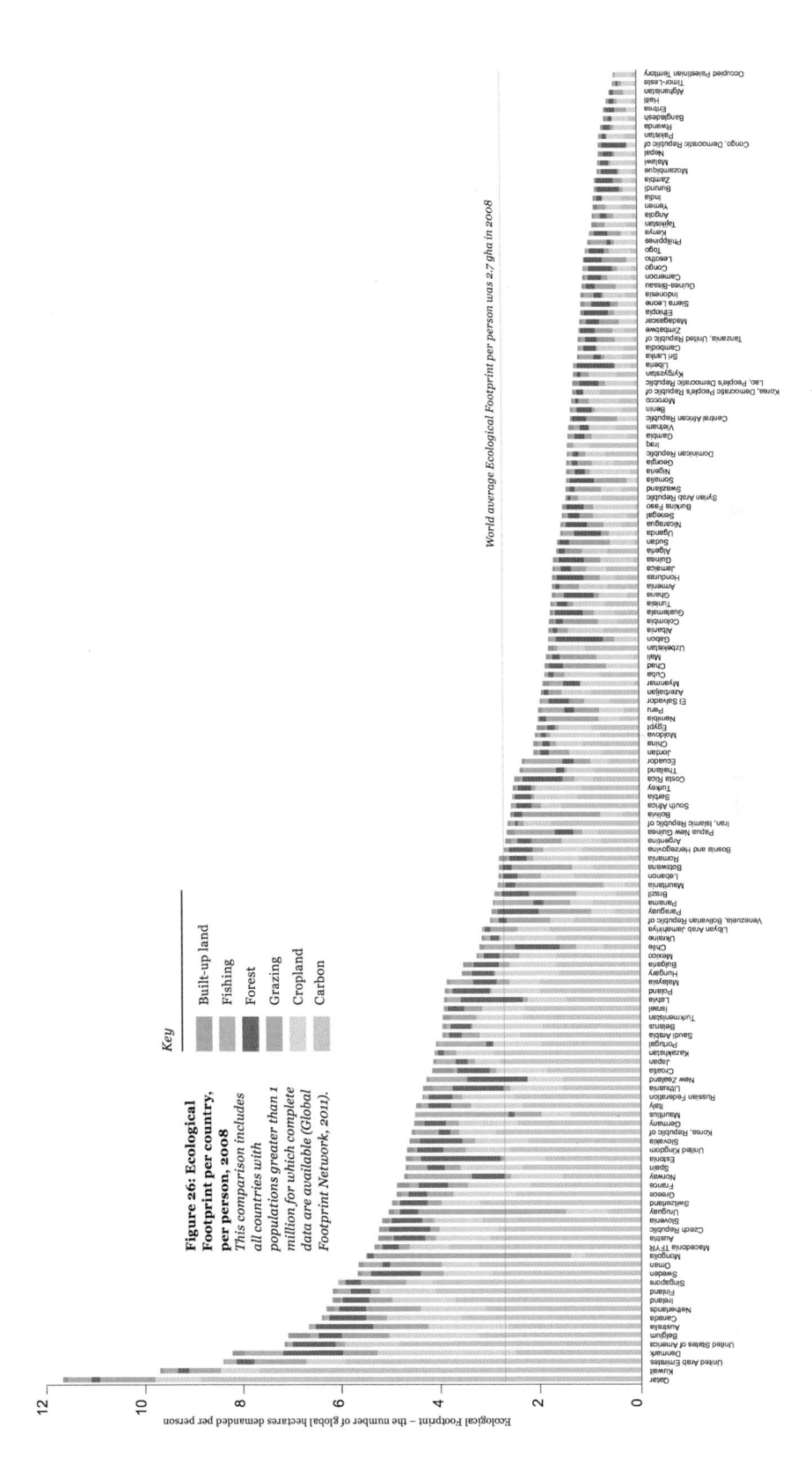

Fig. 10.7 *Ecological footprint per country, per person, 2008 (WWF 2012).*

Comparisons

Images can enable quick assessment of comparative values. Graphs are used to compare quantities. Designers have a wide range of options including the bar chart, pie chart, line chart, area chart, bubble chart, radar chart, histogram, waterfall chart, tree map, scatter plot and box plot. Images are a means to facilitate quantitative comparison to give audiences instantaneous understanding of relative values. Figure 10.7 displays the comparative ecological footprint of nations. The chart illustrates the ecological footprint per capita by country with each line on the graph also divided into different types of natural resources depletion. This bar graph displays vast amounts of data, facilitating comparisons of complex data sets in one image. The ecological footprint is a powerful metric that enables comparison of ecological destruction of different activities in different regions, facilitating analysis and enabling planning towards reducing harmful impacts.

Qualitative Complexity and the Aesthetic Experience

A concern for aesthetics is integral to design. With its focus on aesthetics, design makes artefacts and experiences that feel more pleasurable, meaningful and/or dignified. Good design evokes sensibilities through its look and its feel, that is, perceptual experiences that are associated with beauty and identity. These qualitative features, the domain of the arts, are linked to its aesthetic appeal. This qualitative mode contributes to the potential to engage ecological perception. Laura Sewall writes:

> It has often been said that our environmental crisis is a crisis of perception. We do not see the patterns that would reveal our dependence on the natural world, nor are we commonly aware of the systems within which we are deeply embedded. Our attention, entrained on objects and focused on flat screens, is far removed from the dynamic and animated nonhuman world ... We are as good as blind to the wonder at our feet or the daily spectacle of an ever-changing world. (Sewall 2012, 265)

Images provoke aesthetic experiences in a variety of ways. When designing new work, it is possible to direct audiences to focus attention in ways that encourage particular ways of seeing. Sewall describes shifts in perceptual predispositions. Traditionally, our ways of perceiving are 'conditioned by a cultural history of individualism, mechanism, and reductionion ... [as] ... object focused and cast at the near point'. This myopic vision provokes 'various forms of narcissism' (Ibid., 269). But perceptual habits can be changed. We can move away from this insular modality, towards greater attention to relations, beauty and the numinous. Aesthetics is a means of supporting these shifts.

Due to the way aesthetic experiences trigger notions of beauty, they can be a powerful source of motivation to do the difficult work of creating sustainable futures. Aesthetic experiences have transformative potential. Sewall describes aesthetics in terms of system theory:

> the aesthetic response is a self-organising principle. If we understand ourselves to be self-organizing systems embedded within large systems and as drivers of environmental degradation, a focus on the beautiful view may be the most critical remedy for our current state of disrepair. (2012, 280)

The aesthetic experience is often described as a moment of awe where an individual feels a connection with whatever is provoking the experience. When an observer feels a connection to the object being observed, the stark dualism of the Western tradition is temporarily interrupted. The temporary dissolution of the boundaries of self can trigger new insights, perceptual capacities and pleasure – all simultaneously.

Aesthetic experiences can also be associated with the vitality of a subject, as a subject is captured in a new form and perceived by a participant observer such that 'deep unities' are perceived:

> The aesthetic sense is a capacity to organize meaning in very large quantities of ill-defined information, to detect and create complex relationships and feedback systems, to take into account multiple contexts and frames of reference, and to perceive harmonies and regularities that add up to a deep unity. (Turner 1994 quoted in Sewall 2012, 280)

With these experiences, ecological relations can become an embodied sensation where individuals not only intellectually *understands* ecological context – but have the *experience* of relationship, connection and participation. Feelings of unity, companionship and intimate relations can also inspire empathic sensibilities. These feelings are motivational. Sewall claims:

> In the face of widespread environmental degradation, a relevant and responsible psychology must include whatever inspires and motivates us…an image of desire captures and holds our attention, quickens our pulse, and changes our organic, humming brains. Beyond any rationale, we move: motivated, we act. (Ibid., 280)

If design can inspire desire and pleasure and these feelings can be linked to the empathetic relations with the context that sustains us, then design can function as a catalyst for the emergence of ecological ways of seeing and knowing. Aesthetic experience can be central to the emergence of embodied ecological perception.

Despite their potential, aesthetic experiences cannot be instrumentalized through a design strategy. Moments of insight are emergent phenomenon that cannot be controlled – although they can, potentially, be nurtured. Designers, artists and other cultural creatives can harness their skills for this agenda. The potential for aesthetics responses to provide restorative subsistence should not be underestimated. With these approaches design can help shape our perceptual habits, patterns of behaviour and identities aligned with ecological realities.

11 Ecological Identity

Social practices such as design – and particularly communication design, encourage various aspects of identity. Notions of identity influence the ways in which we treat each other and the environment. Our identities are associated with particular habits of mind that have socio-political-ecological implications. Ecological identity refers to a cultural identity and value system associated with a deeply internalized understanding of ecological connection, interdependence and embeddedness. This sensibility evokes ways of relating that are open, flexible, adaptable and self-aware. Ecological self challenges the boundaries of the reductive ego while situating itself in a network of dynamic relationships. Beyond notions of the isolated self as the basic unit of human experience and in contrast to the neoliberal subject (where social forces emphasis fragmented, competitive and narcissist tendencies) here each self is simultaneously singularly unique and part of the larger social and ecological orders and existing in relation to others. This relational ontology prioritizes of good relations with the ecological and social context.

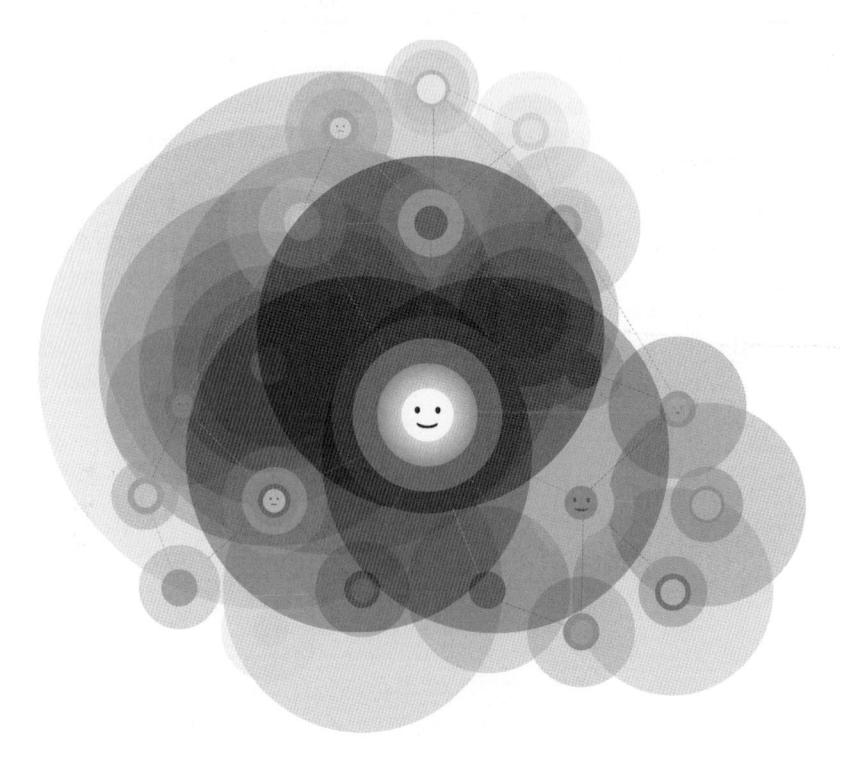

Fig. 11.1 *Ecological Self*. EcoLabs, 2017.

Activating Materialist Values and Narcissists Identities

Certain aspects of identity are associated with environmental problems (wastefulness, selfishness, arrogance, cruelty, etc.). Other aspects of identity foster environmentally beneficial behaviour (conscientiousness, empathy, altruism, cooperativeness). Research described in *The Role of Human Identity* report (Crompton & Kasser 2009) and *The Common Cause Handbook* (Holmes et al. 2011) presents a circumflex model of value systems where values exist in relationship with each other (see Figure 11.2). The psychological research described in these reports indicates that activating values on one side of the circle will suppress values on the other. This work suggests that encouraging extrinsic values (materialistic values) consequently suppresses intrinsic values (personal growth, empathy, community involvement). According to this model, to enable environmental values and behaviour, environmental advocates must seek to diminish those factors that 'prioritise self-enhancing, materialistic values' (Ibid., 35). Particular values support the maintenance of life sustaining conditions over time. Other values are detrimental to this goal. Sustainable transitions depend on the development of the types of values associated with sustainable

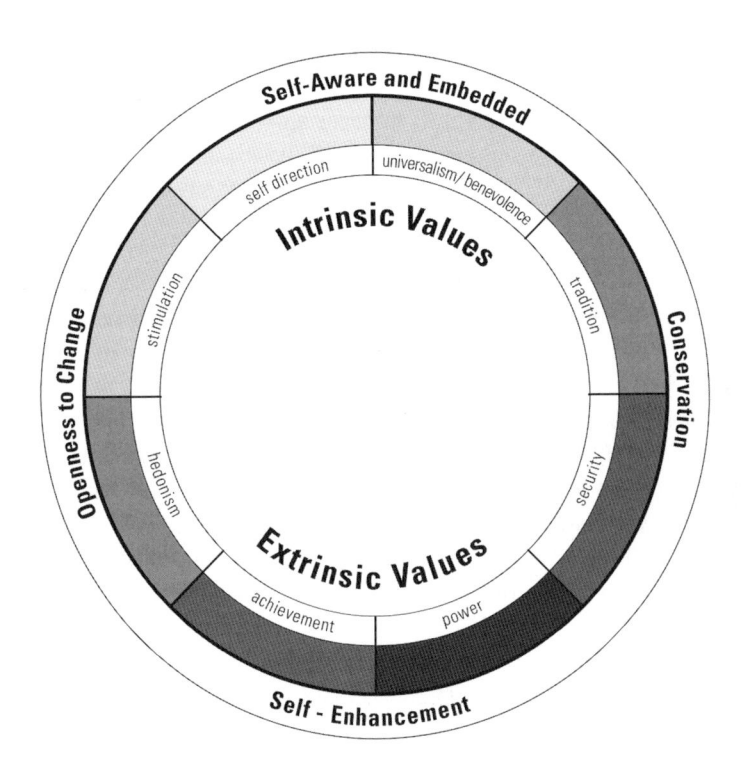

Fig. 11.2 *Circumflex Model of Value Systems* (following Crompton & Kasser 2009; Holmes et al. 2011). I have modified Holmes et al.'s version of this model. EcoLabs, 2017.

behaviours. Behaviour, values and identities are all influenced by social practices such as design. A primary role of design in the development of ecologically oriented futures is the cultivation of the types of idea and identities that will enable ecologically engaged ways of knowing and being.

Communication design can encourage relational, empathic and ethical behaviour – or it can work to stimulate narcissistic behaviour with damaging ecological consequences. Designers typically help cultivate individualistic identities by emphasizing distinctions. In marketing and advertising, designers segregate the public into groups with common consumption habits, prompt individuals to identify with specific target markets and encourage increasingly more specific stylistic preferences. New consumption habits are stimulated as individuals are encouraged to display their uniqueness by identification with various types of clothing, products, gadgets, cars, architecture, graphics and even fonts. Design performs cultural work by encouraging identities aligned with values and sensibilities aligned with capitalism, including those that promote conspicuous consumption. These values are encouraged in advertising and they become embedded in social norms and social relations. Cultural norms change with this process. For example, the once morally objectionable behaviour of narcissism (egotism, vanity, conceit, selfishness) has been removed from the fifth edition of *Diagnostic and Statistical Manual of Mental Disorders* (2013 – known as DSM-5) as an official mental disorder. The trait that was once considered pathological has been reclassified. Since its basic characteristics are nurtured by advertising, mainstream media and corporate culture, this change in what is considered to be morally acceptable behaviour is consistent with its cultural context. Advertising and Reality TV thrive on vanity. Designers encourage consumers to build identities around the products they buy and part of this process involves simulating selfish, narcissist sensibilities.

Cultural Identity Trumps Other Factors

Facts alone are rarely enough to change either beliefs or behaviour within environmental communication or in other politicized areas. The information deficit model, which attributes a lack of understanding to be a result of a lack of information, has been discredited. Tom Crompton explains that 'factual accuracy is an ethical (and practical) imperative ... [but the] presentation of facts alone can even prove to be counterproductive' (2010, 18). The problem lies in the belief that if only people know the facts, they would accurately identify where their self-interest lies and act accordingly. Instead, cognitive scientist George Lakoff claims that if the facts don't support a person's values, 'the facts bounce off' (2004, 17). Herein lies an idea that has profound implications for environmental communications: 'the way that people think, including their response to factual information – tends to work to protect their cultural identity' (Crompton 2010, 19). Cultural identity trumps self-interest, rational analysis and moral reasoning in the way that individuals make sense of communication on politicized issues. Dan Kahan's work on cultural identity and cognition at The Cultural Cognition Project at Yale Law School suggests that

> The prevailing approach is still simply to flood the public with as much sound data as possible on the assumption that the truth is bound, eventually, to drown out its competitors. If, however, the truth carries implication that threatens people's cultural values,

then ... [confronting them with more information] is likely to harden their resistance and increase their willingness to support alternative arguments, no matter how lacking in evidence. (Kahan 2010, 297)

The tendency for individuals to avoid cognitive dissonance (introduced later in this chapter leads to what Kahan refers to as protective cognition, that is the dismissal of facts that threaten identity (2010, 296). The allegiance to ideas that support cultural identity is apparent where people vote against their own class interests (e.g. for candidates that appeal to their sense of identity but do not support their material interests). This analysis implies that new information on the environment is processed in ways that reflect how we conceive of ourselves – our identity.

The limitation of factual information in motivating behaviour change is well known in the advertising industry, public relations and politics. These industries respond with marketing and public relations strategies. Marketers appeal to consumer desire. Political strategists work to deliberately embed particular values in public discourse. Facts are often not enough to change behaviour. Instead, ecologically beneficial values can be nurtured in transparent ways while exposing and challenging anti-environmental biases in advertising, media and institutions.

Strategies of Denial

Many of us have a propensity to avoid uncomfortable information. In *States of Denial* sociologist Stanley Cohen claims that a proclivity to deny disturbing facts is the normal state of affairs for people in an information-saturated society. The book is based on wide-reaching cross-cultural studies including Nazi Germany, South Africa, Israel/Palestine, Rwanda and others zones of human rights abuse, genocide and state-sanctioned violence. Cohen describes strategies of denial on a personal level as psychological and cognitive, and on societal level as communicative and political. Denial functions psychologically below levels of awareness as a 'high speed cognitive mechanism for processing information, like the computer command to delete rather than save' (Cohen 2001, 5). On a cultural level, communication failure and relativistic philosophies reinforce denial strategies.

Cohen's analysis of how disturbing information is avoided is based on violence against people but this work is also relevant for the denial of environmental problems (Marshall 2007, 2009). Denial emerges from complex and multifaceted psychological and cognitive processes. The ability to block out, remain passive, apathetic, indifferent and unresponsive is:

> an unconscious defence mechanism for coping with guilt, anxiety or other disturbing emotions aroused by guilt. The psyche blocks off information that is literally unthinkable or unbearable. The unconscious sets up a barrier that prevents the thought from reaching conscious knowledge. (Cohen 2001, 5)

Denial manifests in three different ways (although each of these has endless characteristics):

1 Literal (nothing happened)
2 Interpretative (what happened is really something else)
3 Implicatory (what happened was justified). (Ibid., 99)

Each of these strategies for denial must be circumvented with care. These types of denial are further complicated by levels at which they become evident:

1. Cognition (not acknowledging the facts)
2. Emotion (not feeling, not being disturbed)
3. Morality (not recognising wrongness or responsibility)
4. Action (not taking steps in response to knowledge). (Ibid., 9)

Each of these states of denial can also be a stage towards acknowledgement as part of a strategy to break denial. For example, where is your own denial situated on the matrix below in regard to any anti-environmental activities or attitude you have? This chart is a tool for thinking about denial. What types of strategies could be used to confront the various types of denial described in this chart?

Confronting Environmental Denial Matrix

	Literal	Interpretative	Implicatory
Cognitive			
Emotional			
Morality			
Action			

States of denial are multiplied across cultural groups where denial becomes institutionalized and systemic. Denial in social groups is exacerbated by communication failure and by powerful interests that benefit from keeping denial working in their favour. On a cultural level, multiple individual cases of denial lead to normalization, then ignoring which develops into collusion: 'people trying to look innocent by not noticing' (Ibid., xii). Denial can lead to cultures of splitting characterized by dissociation and psychic numbing. Cohen describes how cultures plagued by high levels of denial escape into a state that he calls innerism, defined as an 'escape from the public sphere into private life and consumer interests' (Ibid., 156). The spectacles of consumer capitalism clearly provide sufficient distractions. Denial is broken by a process of acknowledgement which 'is what happens to knowledge when it becomes officially sanctioned and enters public discourse' (Ibid., 225). Collective acknowledgement is transformational as it 'makes previously normalized conditions into social problems… [and then]…social institutions, policy strategies, and, even a new language are in place to undermine denial and encourage and channel individual acknowledgement' (Ibid., 250). This change comes about through the work of social actors and social movements that chip away and eventually shatter denial.[1] Strategies for breaking denial involve nurturing individual and social capacities for acceptance of historical circumstances. Cohen's work suggests that we must aim beyond merely presenting the facts and invoking moral arguments to addressing the social and psychological mechanisms that support denial. Enabling new agencies is part of this work.

For those who are already well aware of ecological circumstances learning to accept denial as a common phenomenon and not demonize those who perpetuate denial (despite the destructive consequences that are a result of the lack of political will and motivation for change) gives others the space to learn. Becoming capable of making necessary changes in perspective and action is not easy. The psychological difficulties associated with acknowledging and responding to disturbing environmental information need to be taken seriously. Recognizing denial as commonplace can be a starting point:

> Instead of agonizing about why denial occurs, we should take this state for granted. The theoretical question is not 'why do we shut out?' but 'what do we ever not shut out?' The empirical problem is not to uncover yet ever more evidence of denial, but to discover the conditions under which information is acknowledged and acted upon. The political problem is how to create these conditions. (Ibid., 249)

Design can work strategically to end the normalization of denial by helping to create conditions for collective acknowledgement. Recognizing the ways that denial manifests is a step in the right direction. The two illustrations below illustrate emotional responses to disturbing environmental news (Figures 11.3 and 11.4). Since the consequences of ecological harm deny others the right to a healthy environment, a condition we all need to live, denial of ecological circumstances can be framed as a category of social deviance: a pathology. The lessons learned from struggles against human rights abuses inform anti-denial work for the environment.

Cognitive Dissonance

Cognitive dissonance refers to the theory that people will avoid holding conflicting ideas simultaneously due to uncomfortable feelings of dissonance that arise. To avoid dissonance we deny information that causes the conflict. Cognitive dissonance is a central problem in environmental communication due to the discomfort that arises when encountering disconcerting environmental information. Accepting that our consumption habits are a cause of

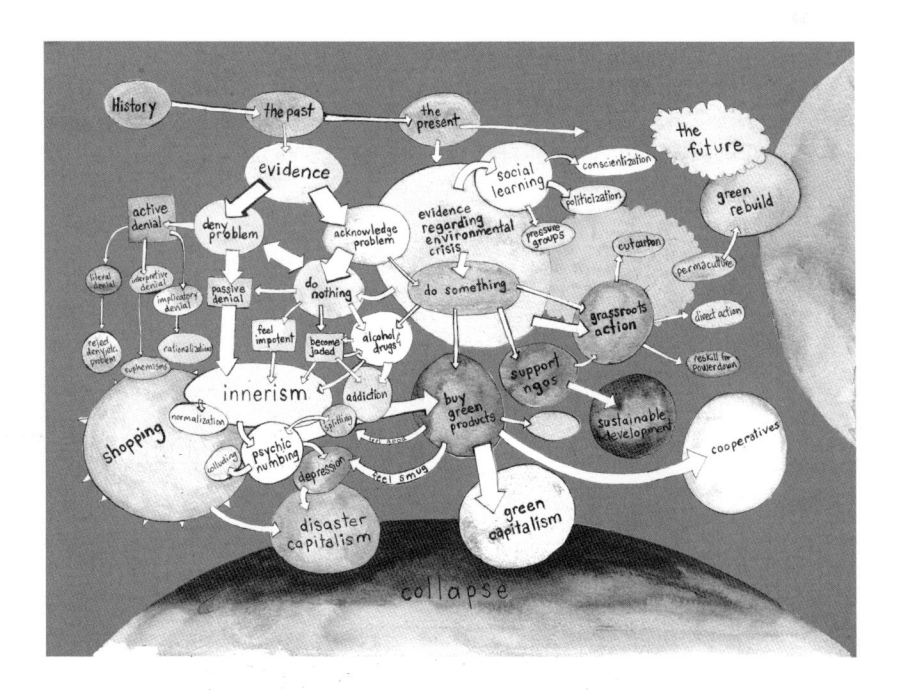

Fig. 11.3 *The Psychology of Environmental Crisis*: *Denial Bubbles*. EcoLabs, 2017.

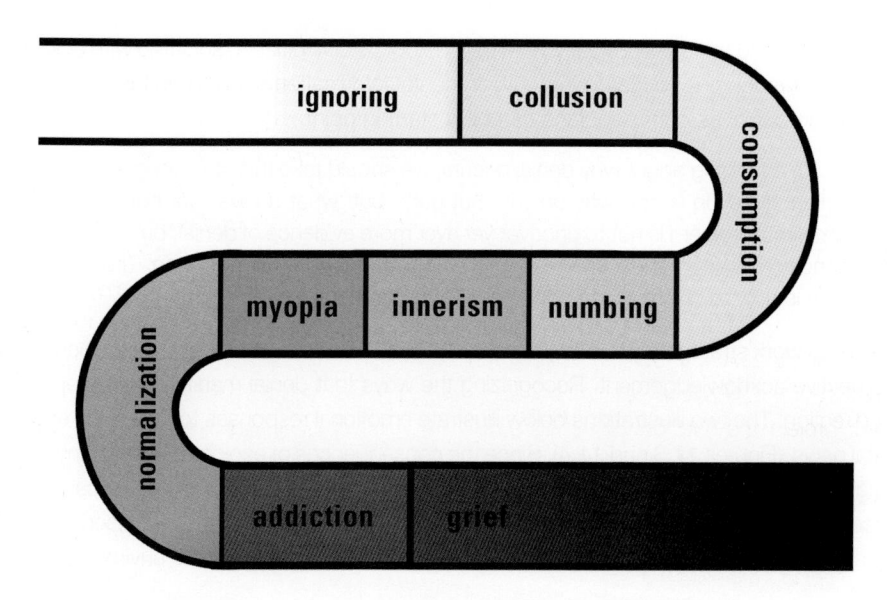

Fig. 11.4 *The Psychology of Environmental Crisis: Trajectory of Denial.* EcoLabs, 2017.

the ecological crisis conflicts with the desire to conceive of oneself as essentially benign and thus cognitive dissonance triggers various forms of denial. It is difficult to accept the reality of atmospheric deterioration and other environmental problems if one is a willing participant in consumer culture. The contradiction between self-image as basically good and the awareness that the lifestyle choices and the decisions we make professionally and politically are perpetuating ecocide is psychologically difficult. For most people with consumer lifestyles, cognitive dissonance is likely a standard psychological trait. Defence mechanisms create a heavy armour of resistance to disturbing information. The more powerful actors are also those with the highest carbon emissions and ecological footprint per capita. With more power comes more responsibility and (arguably) higher degree of cognitive dissonance. The theory suggests that coercive and didactic communication strategies can provoke an opposite effect because they trigger these strong defensive psychological responses. One key insight is that acknowledgement often comes along with pro-environmental action since the act of doing something (even small acts) lessens cognitive dissonance as we move, ever slightly, away from an identity that creates dissonance. Ecopsychology offers more responsive strategies.

Ecopsychology: The Ecological + the Psychological

Ecopsychology is the study of the psychology of human–nature relations. Ecopsychology theorizes the ways in which we understand ourselves as ecological beings and provides therapeutic and communicative practices to engage ecological identity. The term was coined in 1992 by cultural historian Theodore Roszak as an emerging focus on industrial culture's 'longstanding, historical gulf between the psychological and the ecological' (Roszak 1992, 14). Ecopsychology aims to create 'the subjective conditions for an ecological society' by offering 'a corrective to anthropocentrism in psychology' while also 'questioning of our entire social formation insofar as it generates ecological and psychospiritual crisis through domination of nature' (Fisher 2012, 100, 81–82). It investigates the relationship between the ecological and the psychological within the context of severe environmental degradation.

Ecopsychologists theorize a conflict arising from ontological ecological embeddedness and assumptions of radical autonomy (epistemological error). This internal conflict results in a variety of psychological problems. The denial of ecological self is theorized as at the root of the alienation and pathologies of the current era. We are all embedded within and part of the Earth (whether in the city or in the woods) but our ideas, sensations and emotions often do not reflect this reality. This conflict takes a severe psychological toll. The repression of ecological unconscious is as much, if not more, of a source of psychological dysfunction as the repression of innate sexual impulses, as first theorized by Sigmund Freud (Roszak 1992, 1995; Metzer 1995, 62). Psychotherapists believe that it requires 'vast amounts of energy to repress and/or project the shadow' (Harper 1995, 194). The repression of the ecological self and connection to the non-human natural world is seen as profoundly damaging to psychological well-being. Breaking through this pathology is part of the process of enabling new ways of relating to the Earth and each other – paving the way for the emergence of ecological self.

Neologisms for Psychoterratic States

Ecological theorists and ecopsychologists propose a maturing process that involves a gradual broadening of one's identifications, opening the possibility for deeper affinity for non-human nature. Conceptual resources to enable this process include neologisms to describe psychological conditions. The idea of 'biophilia' was first coined by Erick Fromme (1964) but was popularized and brought into biology and conservation by E.O. Wilson (1984). It refers to an emotional affiliation with non-human nature that is considered to be innate in humans but repressed by modern industrial culture. The Nature Deficit Disorder hypothesis (Richard Louv 2006) suggests that children and adults without experiences either in nature or even outdoors can suffer from a range of psychological problems. Philosopher Glenn Albrecht has coined several terms to describe specific emotional responses to ecological circumstances and relationships as 'a conceptual framework for understanding psychoterratic, or earth related (terra) mental health (psyche) states or conditions' (2013). Below is a list of various earth related mental states (as proposed by Albrecht and others):

Psychoterratic States
Solastalgia: a feeling of desolation of the deterioration of a loved home.
Soliphilia: the love of nature and the solidarity needed to resist forces of desolation and restore environments
Topophilia: the love a particular place, or landscape in general
Ecoanxiety: anxiety based on ecological conditions.
Eutierria: a feeling of elation or oneness with the Earth
Ecoparalysis: the inability, due to fear or hopelessness, to respond to environmental problems. Not necessarily an expression of apathy.
Ecophobia and *ecoaversion*: aversion to the environment. Considered to be an effect of socialization where experiences in nature are absent or negative. (Albrecht 2012, 251–256, 2013)

These neologisms enable richer understanding of human relations with non-human nature. Albrecht's investigation of psychological states suggests that the lack of action on environmental issues could have psychoterratic origins rather than any active decision to do nothing (2012, 258). Using the concepts listed above, designers can explore psychoterratic experiences and consider how they might impact design problems and potential solutions.

Mental Health, Sanity and Additions

Ecopsychology reflects on what constitutes sanity. It suggests that once psychology acknowledges the ecological dimension, a new definition of psychological health emerges. Ecopsychology proposes an expanded definition of sanity and 'an environmentally based criterion for mental health' (Roszak 1995, 15). In *Nature and Madness* (1982), Paul Sheppard describes a 'psychopathological metaphor for our relationship with the natural world' (Metzer 1995, 56). Sanity is dependent on 'relationships with your environment' (Gray 1995, 173). By failing to acknowledge ecological context, traditional psychology perpetuates dysfunction and current pathologies. To illustrate risks associated with traditional psychological practice, Terrance O'Connor tells the true story of a woman in Nazi Germany who spent three years in therapy to overcome 'irrational fears'. A few weeks after her treatment ended, the woman (who was Jewish) was shipped to a concentration camp. O'Connor asks his fellow psychologists: 'By helping people adapt to a destructive society, are we doing more harm than good?' (1995, 150). Ecopsychologists reject the idea that the normative state in a culture that is destroying its habitat is psychologically healthy (since it is maladaptive to normalize behaviour that is ultimately self-destructive). This reconceptualization of sanity and mental health can play a powerful role in the struggle to legitimize and normalize pro-environmental values and behaviour. Part of this works involves politicizing mental health.

Some ecological theorists and ecopsychologists theorize that damage done to the Earth as self-harm (Bateson 1972; Macy 1995; O'Connor 1995). Certain people are especially sensitive and they feel the ecological crisis as irrepressible pain. This grief is especially acute in spaces where environmental disasters have occurred (i.e. Bhopal disaster 1984; Chernobyl 1986; Deepwater Horizon 2010). Herein lies a core difficulty within the environmental movement. On the one hand, ecopsychologists hold that 'for an environmental ethic to succeed, nature needs to be meaningful to us on a variety of levels, including the emotional' (Windle 1995, 136). Yet becoming emotionally aware of ecological embeddedness is also becoming emotionally sensitive to the damage we are doing to ourselves, other humans, animals, natural spaces, and so on, as part of an ecologically destructive society. Acknowledging ourselves as embedded on an emotional level as well as a cognitive one, in an environment that we are currently in the process of severely damaging, is often a psychologically disturbing and even painful process. This despair is a site of attention for ecopsychologists. Joanna Macy facilitates 'the work that reconnects' with front-line communities suffering environmental devastation, accompanying illnesses and mortalities (such as in the towns surrounding Chernobyl), as well as with other people suffering solastalgia, ecoanxiety or depression related to the environment.

Some ecopsychologists use the metaphor of addiction to describe environmentally destructive behaviour. Ralph Metzer claims that 'our inability to stop suicidal and ecocidal behaviour fits the clinical definition of addiction or compulsion: behaviour that continues in spite of the individual knowing that it is destructive to self, family, work and social relationships' (1995, 60). Environmentally destructive behaviour and compulsions arise from dissociative alienation and pathologies rooted in the repression of ecological unconscious.

Ecopsychology characterizes these states as a narrowing of awareness, anesthetization (Roszak 1995) or a psychic numbing (Spratnek 1997, 77). To break these patterns of thought, something affirmative must provide an alternative.

Confronting Disturbing Information

Expecting individuals to change deep-seated patterns of thought by simply presenting them with the facts is often unproductive. From the ecopsychology perspective, it is obvious that information alone will not work towards changing behaviours or attitudes in regard to the environment:

> Terrifying information about the effects of environmental destruction can drive us deeper into denial and feelings of futility, unless we deal with the responses it arouses in us. We need to deal with this information on a psychological and emotional level in order to fully respond on the cognitive level. We already know we are in danger. The essential question is: can we free ourselves to respond? (Macy 1995, 252)

Environmental information can provoke similar defensive reactions to those encountered by therapists working with drugs/alcohol addiction. The addiction diagnosis indicates that guilt and shame can make things worse. Ecopsychology suggests that the environmental movement has overused 'shame and blame' strategies such that the public is defensive, rigid, numb and overwhelmed (Kanner & Gomes 1995; Rozsak 1995). Environmental educators, environmental communicators and designers can respond with more subtle communication strategies.

> Although our inner lives are relentlessly diminished by ecosocial isolation, the antidote lies in recovering awareness of our context. We are embodied and embedded in a dynamic sphere of physical relationships and processes that create real commonalities, which have been denied by ideologies of both the rugged individual and the fragmenting 'politics of identity'. (Spretnak 1997, 8)

Social and dialogic learning processes (with approaches such as critical ecopedagogy) support ecological learning. With these methods individuals process how they feel when confronting disturbing information. These practices are a means of nurturing new agencies to act as they help individuals deal with the emotional and motivational implications of encountering environmental information.

Ecopsychology describes a variety of psychological processes involved with confronting and responding to environmental challenges. Some theorists make a distinction between 'environmental' and 'ecological' approaches. Andy Fisher describes environmental approach as technocratic: 'when nature or the environment is regarded as external to us, we are simply technical managers or the biosphere' (2012, 93). The ecological approach understands each individual as embedded and thus the psychology of internal relations becomes relevant. For ecopsychologists, since the study of the human condition involves human subjectivities, the ecological approach necessary.

Gregory Bateson's ecological and psychiatric exploration of the roots of mental illness in *Steps to an Ecology of Mind* (based largely on his work with mental patients, studies of alcoholism and schizophrenia) was foundational to the development of both ecological theory and ecopsychology. Bateson first called for the merging of the psychological, social and ecological spheres: a feat that can only be accomplished by moving beyond the limitations

Fig. 11.5 *We Are Nature Protecting Itself*. John Jordan, Isabelle Fremaux, Selcuk Balamir and associates at The Laboratory of Insurrectionary Imagination, Climate Games #HackCop21.

of a reductionist model of the human psyche and knowledge systems. Ecopsychology supports this expansive work at the intersection of ecology and psychology by theorizing this space and also providing practical methods to deal with profound shifts in our understanding of human–nature relations. It explores the emotional difficulties of confronting civilization-threatening environmental problems.

Ecopsychologists often use stories to help with this work. Joanna Macy, Chris Johnston and David Korten describes the story of a contemporary Great Unraveling where 'when profligate consumption exceeded Earth's capacity to sustain and led to an accelerating wave of collapsing environmental systems, violent competition for what remained of the planet's resources, and a dramatic dieback of the human population?' (Korten 2006b). The story of The Great Turning is a narrative that helps some avoid succumbing to either panic or paralysis in order to do the work necessary to enable a future life-sustaining civilization (Korten 2006; Macy and Johnston 2016). Narratives such these are pathways to navigate difficult times, to help individuals embrace change and acknowledge ecological identity. Other storytellers who engage the ecological imagination and identity are artists such as John Jordan and the Laboratory of Insurrectionary Imagination (Labofii) art collective. The collective worked with designer Selçuk Balamir and others towards the spectacular *COP21 Climate Games* at the UN summit in Paris in December 2015. Their work is explicit in the artists' identification with the ecological. Their signs and stickers read: 'We are not fighting for nature. We are nature protecting itself' (see Figure 11.5).

Note

1. Cohen notes that the longest lasting anti-denial movements have all been started and sustained by women (2001, 95).

Part 3 Politics

Part 3 Polities

12 Social Marketing

Social marketing is defined by UK's National Social Marketing Centre (NSMC) as 'a proven tool for influencing behaviour in a sustainable and cost-effective way' (2015). The NSMC claims that social marketing is not about changing what people think or how aware they are about an issue, but about behaviour. But behaviour is known to emerge from individual values and attitudes (which we learn). In practice, social marketing is the method through which corporations respond to social and environmental issues by marketing themselves socially responsible. It is the advertising industry's response to growing public anxiety about converging eco-social crises. Since creating an image of corporate social responsibility is easier than doing the work transforming unsustainable corporate practice, the regulatory framework or the economic model itself, social marketing is often seen as good practice. Ideally, the image of corporate social responsibility presented by social marketing campaigns in reflected in practice, but this is not always evident. In this chapter, I use the Hopenhagen campaign at the 2009 Climate Summit in Copenhagen (UNFCC, Cop-15) as an example of how social marketing can derail the social and political change urgently required to address environmental problems. I suggest alternative strategies for the design of public information campaigns on issues of controversy and propose that it is wrong to assume that tactics used in the advertising industry can be used to support effective communication and design strategies on social and environmental issues.

Social Marketing Is a Distinct Type of Communication

Environmental problems require a deeper engagement with pro-environmental values than social marketing practices typically facilitate. Social marketing never truly prioritize environmentally sound alternatives because its frame of reference occurs entirely within the parameters and priorities of the market. Market forces are always the determining factor in its analysis and practice. While social marketing presents itself as against greenwashing, its practitioners do not link exploitation and pollution to systemic processes in the economic system and certainly not to the brands that hire social marketers. As a powerful discursive practice social marketing often functions to divert attention *away* from the drivers of conspicuous conception. It thereby becomes a primary tool for corporate interests who wish to maintain the status quo while appearing to do the right thing by the environment.

Social marketing engages with social and environmental issues in ways that do not challenge the basic premises and priorities of the market. It cannot support proportional responses on problems such as climate change because marketers 'fail to question the dominance of today's individualistic and materialistic values' (Crompton 2008, 5) in the material they produce for audiences (now called 'consumers'). The danger is that 'environmental challenges will not be met while we maintain a narrow focus on the happy coincidence of self-interest and environmental prudence' (Ibid, 2). Furthermore,

> The marketing approach to promulgating behavioural change…is doubtless effective at generating piecemeal change where this is at its most painless – particularly where such change is embodied in the purchase of a new product. But in the course of embracing the more systemic and structural changes that are needed they may be at best a distraction, and at worst a procrastination. (Ibid., 8)

The practice assumes that extrinsic values (i.e. acquisition of material goods, financial success, image and social recognition) can be linked to social and environmental values. By emphasizing new consumer habits as the route to sustainability this practice reinforces the sovereignty of consumer choice over other priorities. This logic determines that social marketing is bound in contradictions (since sustainability requires less consumption of resources but the market requires market growth, based on more consumption). Within this context, slightly better consumption habits are dwarfed by dramatic increases in gross consumption.

Advocates of social marketing argue that values arise from human nature and thus conspicuous consumption is an inbuilt feature of human psychology.[1] These arguments are disingenuous about the powerful practice of marketing itself. If marketing was not so enormously influential it would not be the trillion-dollar industry that it is. People internalize values that are part of the cultural environment (often uncritically). Attitudes and values are shaped by culture regardless of whether or not we are conscious of these processes or not. Communication practices serve to activate values, so it is worth noting what values different practices engage. There are alternatives. Anthropological, sociological and historical knowledge about values systems, political systems and social change documented by scholars such as sociologist Marcel Mauss (1925), political scientist Elinor Ostrom (1990) and anthropologist David Graeber (2000, 2011) describe alternative value systems emerging from different ways of organizing social relations and the distribution of resources. Values are developed by socialization. Design is a practice that encourages various social norms and values. Social marketing typically nurtures the values that suit the interests of its clients first and society at large and the environment second (if at all).

The Case of Hopenhagen

Hopenhagen was a social marketing initiative launched by the United Nations and the International Advertising Association leading up to and during the United Nations Climate Change Conference (COP15) in Copenhagen December 2009. The massive international advertising and public relations campaign presented itself as a people's movement to stop climate change. It culminated with a huge installation in the public square in central Copenhagen during the COP15 climate summit. In the midst of a heavily contested political space, Hopenhagen appeared to be a people's movement supported by corporate sponsors who were helping governments save the world. The campaign demonstrated the dangers of conflating marketing with social change and illustrated tensions within the practice of social marketing itself. This conflict of interests became evident when climate activists in Copenhagen made the Hopenhagen campaign itself a target of their protests.

The Hopenhagen project was initiated at Davos in January 2009 when UN Secretary-General Ban Ki-moon asked for help from the International Advertising Association (IAA) to build awareness and public support for a strong deal at the Copenhagen Climate Summit later that year. The Kyoto Accord was due to be replaced and Copenhagen was an opportunity to finally make a binding global agreement. Ban Ki-moon had high hopes for Copenhagen and called on the world's advertising industry to help:

> to create a strategy to harness all the brilliance, innovation and creativity that the marketing industry is known for…it should be THE climate change communication initiative. We hope it will be a game-changer. It will explain, educate and ask for global engagement leading to success in Copenhagen. (2009)

The conflicts of interests that would inevitably be played out over the following year should have been obvious. The campaign was produced by the IAA and Ogilvy Earth with an assortment of large advertising and public relation firms. Oglivy Earth had been establishing itself as a centre of expertise for avoiding greenwash[2] and helping corporate clients brand themselves as good environmental citizens. Most famously they helped BP rebrand as 'Beyond Petroleum' in 2000 with a $200 million rebrand campaign (SourceWatch 2011). The Deepwater Horizon Oil spill in 2010 released 4.9 million barrels of real crude oil into the Gulf demonstrating that BP is not actually 'beyond' petroleum at all. Nevertheless, Oglivy maintains that brands need the Oglivy 'greenwash compass'.

Hopenhagen was unveiled at the Cannes Lions International Advertising Festival in June 2009 then at New York's JFK International, LA International and London Heathrow airports and followed by an international campaign that included a website and petition where six million people signed up and became 'citizens of Hopenhagen'. Hopenhagen had a massive outdoor presence on gigantic billboards around Copenhagen during the COP15 Summit. The central focus was an installation in a downtown square featuring sustainability innovations and corporate advertising. A giant globe displayed projection updates on the progress of the summit. Exhibits demonstrated green technologies. An observer who was not aware of the political struggles in the climate movement might be led to believe that Hopenhagen represented the interests of thousands of grassroots activists who had been organizing for COP15 and had travelled to Copenhagen for campaigning, demonstrations and direct action.

The campaign created significant opportunities for sponsors (Coke, SAP, Siemens, Gap, BMW and DuPont and others) to demonstrate their goodwill as corporate citizens. Towering billboards were plastered around Copenhagen including a giant illustrated Coke advert: 'A Bottle of Hope' (Figure 12.1) conflated the work of social movements with the commercial aspirations of Coca-Cola. The cross-media spectacle of Hopenhagen projected the image of being a movement against climate change wherein people, the government and corporations were all working together in harmony with Coke, Siemens, BMW, etc. as part of the climate movement. The stark differences between the campaign messaging and reality were apparent in reports from the negotiations. In the context of a deadlock created by what many describe as the corporate capture of the climate negotiation, Hopenhagen itself became the object of scorn, derision and protests.

To climate justice activists (such as Climate Camp UK, Climate Justice Action and Climate Justice Now!) Hopenhagen came to represent the corporate takeover of public space, people's movements and political processes. Activists argue that as corporations attempt to align themselves with the creativity, passion, energy and moral force of social movements the work of the movements themselves is obscured. A Corporate Watch report describes how:

Fig. 12.1 Hopenhagen advert: A Bottle of Hope. Photograph by author, 2009.

'The case of "Hopenhagen" illustrates how the triad of Corporation – PR Company – and NGO operates to create and then "engage" with the "citizen consumer" in the wake of effective global critiques and boycott movements' (Schling 2011, 11). Environmental problems and protests present a significant threat to claims of corporate social responsibility (CSR). Social marketing is a primary strategy harnessed by corporations to mitigate these challenges to their legitimacy. In this case, Hopenhagen appropriated the creativity and energy of the climate movement to serve the very interests that the movement was attempting to expose as those most responsible for climate change. Hopenhagen worked as a high-profile campaign that made it appear (to the disengaged) as if the campaigns sponsors were on the side of the climate movement. Meanwhile, some activists (including The Yes Men, Climate Camp UK and independent activists) found Hopenhagen so offensive that they made the campaign itself an object of their protests with subvertising, direct action, banner drops and an occupation with tents one cold December night at the central Hopenhagen installation (see Figure 12.2).

Fig. 12.2 Hopenhagen installation in central Copenhagen and Climate Camp occupation. Photos by Kristian Buus, ©kbuus, 2009.

Don't Corporations Also Want to 'Save the World'?

Individuals within many corporations undoubtedly do want to 'save the world'. It is highly unlikely they will so as long as their efforts remain fixated the profit margins of their employers. For example, while the explicit aim of the Hopenhagen campaign was to build a movement to stop climate change, the implicit aim was to cast corporate sponsors as good corporate citizens. Apparently Hopenhagen was created by many advertising employees working on a *pro bono* basis. But the contradictions between explicit messaging and implicit intentions increase cynicism and generate divisive tensions – and so this *pro-bono* work is not benign. Protests against Hopenhagen illustrate the conflicts of interests evident in the spectacle.

The case of Hopenhagen raises several questions. Why would anyone at the UN expect the advertising industry to be capable of operating as 'a game-changer' (Ban 2009) on environmental issues? Why would the UN expect the International Advertising Association to transcend its own ideological commitments? The answer is a fundamental confusion about social communication and the lack of critical thinking on issues of communication, design and marketing. The UN asked the IAA to help with COP15 because they conflate advertising with communication. The marketplace has become such a dominant mechanism for mediating social relations that marketing is considered by some to be akin to social communication. This is a dangerous naturalism. The market is not the arbitrator of all values, nor is it ideologically neutral. People are more than mere 'consumers'. Communication campaigns that are designed as social marketing embed market priorities and ideological assumptions into the messaging. They fail to disturb the priorities of corporate capitalism and thereby contribute to the business as usual narrative.

Ban Ki-Moon thought the adverting industry could educate and motivate the public about climate change. In fact, the advertising industry is keenly aware of their corporate clients' interests and will communicate the agendas and discourses that suit these interests even when they try to address the issue of climate change. So, Hopenhagen was not actually about creating systemic change necessary to address the climate crisis: but it was a brilliant marketing opportunity for Coca-Cola and other big brands. This analysis is not a matter of liking or disliking corporations. It is about thinking critically about conflicts of interest and corporate practice. A strong deal at Copenhagen or subsequent climate conferences demands legally binding commitments by global governments to keep emissions below 350 parts per million with an equitable distribution of carbon allocations across global populations. A just deal would acknowledge historic responsibility for carbon emissions. These are the political problems faced by negotiators. It is disingenuous and anti-democratic to create a communication campaign that ignores these debates. Hopenhagen used the skills of designers to create a spectacle of corporate concern with no actual agenda for change on the order that would address climate change. Corporations claim to want to address climate change but simultaneously work to erode capacities for regulation and undermine democratic processes. The failure to reach a strong agreement at Copenhagen was a (short-sighted and ill-informed) 'victory' for corporations as it enables them to continue avoid regulation.

Lessons for Environmental Communication and Campaigns

The case of Hopenhagen demonstrates how social marketing erodes the political demands of social movements. As such it displays contradictions within the concepts of social marketing.

Marketing typically functions as a driver of consumption. Activists struggle to combat the forces of globalization (anti-democratic processes, corporate enclosures, privatization, corruption and ecological devastation) by forming social movements and resisting the corporatization of the commons and everyday life. Marketers are employed by corporations concerned with the creation of profit while projecting an image of responsibility. These two forces are integrally at odds. The fact that this conflict of interest is not acknowledged demonstrates how well capitalism has succeeded in naturalizing its own profit-seeking agenda. What lessons for communication and design can be drawn from the example of Hopenhagen?

ENABLE DEMOCRACY AND DEBATE

Climate change and other environmental issues are political problems. Public deliberation is central to the democratic processes needed to work towards solutions. Communication campaigns need to be more than top-down messaging and public relations strategies broadcasting corporate-friendly messages. Confronting environmental realities is a learning process. Public exhibitions and campaigns must reflect the nature of contested information and provide space for debate. Where marketing practices typically approach audiences as passive consumers, communication design on issues of the environment must encourage active participation in processes of learning and change-making. In attempting to create the illusion of a bottom-up initiative, Hopenhagen was designed as an idealized vision of what a climate movement might look like. What was missing was any sense of the intense conflicts, the power struggles and the contested nature of the policy opinions.

RESIST DEPOLITIZATION AND REVEAL DISCURSIVE PLURALITIES

The communication of heavily contested material requires a recognition of a variety of political agendas and constituencies. Hopenhagen undermined the potential for social change by co-opting the language and imagery of climate movements and then obscuring their political demands. In doing so, Hopenhagen generated conflict between environmental movements and those who wanted to harness the energy of these movements to boost their own brand. The Hopenhagen campaign stated; '6.8 Billion People – One Voice. Together we can fix climate change'. But six billion people do not speak with the same voice as Hopenhagen's platform. In fact, many of the people most concerned about climate change and attentive to the lack of progress being made in the negotiations were the climate activists themselves who were tear gassed, pepper strayed and arrested by police during marches and protests over the two-week period. Inside the conference halls official delegates claimed progress was stalled by the corporate lobby. Many delegates left the conference hall to join the demonstrations once it became apparent that there was no intention by the main negotiation teams to make binding commitments. Outside protesters and delegates were greeted by the police and tear gas. Meanwhile, Hopenhagen erased the diverse voices at Copenhagen to make its own corporate-friendly agenda appear as the universal voice: the 'One Voice'. Revealing discursive pluralities is necessary on issues of controversy such as the environment.

CHALLENGE GREENWASHING AND MISINFORMATION

Information about complex issues such as climate change needs to reflect the fact that topics such as bio-fuels, air travel and carbon trading are highly contested issues. It is disingenuous to ignore these debate. Hopenhagen was a high-profile example of the precarious path environmental campaigns take between generating excitement around the issues and generating material that masks unsound environmental practice and supports uncritical (but corporate-friendly) points of view. Unfounded technological claims (for geoengineering, for example) dislodge the work of changing consumption habits to reduce carbon emissions and the pressure for political change. Corporations claim new products will solve environmental problems but the evidence does not support these claims. Hopenhagen promoted the popular fiction that climate change can be avoided through intelligent shopping. Campaigners have good reasons to believe that this approach will not work and seek political solutions. When corporations determine which information forms the basis for public discourse, knowledge itself is dangerously distorted. The environmental justice movement fiercely contests this discourse.

Good Intentions Are Not Enough

Designing environmental campaigns and public exhibitions on politicized issues of the environment demands different techniques than those in the social marketing toolkit. Environmental problems require engagement with the complexities and core assumptions that drive unsustainable ways of living. There is a wealth of experience within social movements, academic research and professional practice to facilitate environmental learning. A well-designed environmental communications campaign must engage with the complexity of relevant controversies. Environmental communication should function to connect science, policy and the public in ways that allow public ownership of the debate. The advertising industry is out of its depth in this area. Sustainable futures depend not only on a greater level of awareness but also on a critical mass of individuals with emergent capacities for enacting personal and political change. The gaps between values and abilities or willingness to change behaviour and political alliances based on these values are encouraged, enabled or disabled by communicative practices. Environmental communication campaigns must function to help individuals find a sense of agency in the face of change. This is will not happen with advertising practices that work in a top-down manner sending messages to consumers about the good deeds and intentions of corporations.

Hopenhagen is a classic example of the corporate appropriation the creativity, passion and moral energy of people's movements and the subsequent neutralization of the messages demanding structural change and climate justice. The campaign confused and derailed the climate movement to serve commercial interests of Coke, Siemens, Gap, BMW and other participating brands. Hopenhagen might have been conceived with good intentions by people working on a pro bono basis, but it is wrong to assume that the tactics used in the advertising industry can be used to build a climate movement. Good intentions are not enough. The non-result in Copenhagen was a hollow victory for corporations who are resisting regulation. It was a devastating failure for those who take the dangers of climate change seriously enough to organize towards dramatic reduction of greenhouse gas emissions.

Notes

1. See debates between Solitaire Townsend from social marketing consultancy Futerra and Tom Crompton (2010).
2. Oglivy Earth published a 'From Greenwash to Great' report in 2010.

13 The Green Economy

The ways in which we conceptualize the environment influence both policy and design. They under girth political decision-making and establish how we care for the space we inhabit. The emerging green economy project is a good example of how concepts about the environment are constructed, how they can become naturalized as basic assumptions and how they can then be used to design the political mechanisms that are used to address environmental problems. The green economy project was launched June 2012 at Earth Summit Rio+20 (United Nations Conference on Sustainable Development, UNCSD), by the United Nations Environment Program (UNEP). The project presents itself as a far-reaching programme of reform. While the rhetoric suggests that the UN is serious about addressing the biodiversity crisis, a wide variety of social movements are less convinced by the proposed policy mechanisms. At Rio+20 civil society responded with a plethora of critical responses. The UNEP's green economy uses environmental language to advance proposals that are revisioning the processes through which the global commons will be conceptualized and managed. Some of these proposals are antithetical to many of the environmental movement's basic premises as I will explain below.

The UNEP's green economy project (GEP) is part of a political project wherein environmental decision-making is shifting from political processes to economistic processes and logic. With this change, environmental concerns, concepts and language are used to describe policy agendas where markets determine environmental priorities. Despite the green rhetoric, critics claim that this version of the green economy will give more control and power to the corporate and the financial sectors to manage nature through natural capital accounting processes. Thus in stark opposition to what green economists have traditionally conceived of as the green economy, with its emphasis on democratic decision-making, the project moves environmental decision-making out of the public sphere and into the marketplace. Decisions about the environment will be made by those with the capital to participate in these markets.

While most scientists and environmentalists involved with the GEP aim to find a means of helping/making political and economic actors acknowledge the value of the environment, submitting nature to the logic of the market and the financial industry is a risky enterprise. Instead, green and ecological economists maintain that the economic system must be designed to respect the needs of ecological systems that provide the geophysical context for economic systems to exist in the first place. By using the term 'green economy' the project obscures the differences between different ways of managing the ecological commons, that is, a natural commons-based approach[1] versus a natural capital-based approach.[2] The project promises to provide economic mechanisms to stop the destruction of natural spaces and species, but there are good reasons to be sceptical.

The GEP aims to protect nature by accounting for 'externalities' of environmental damage. Environmental externalities are consequences of economic processes that affect people not involved in the economic transaction and who did not choose to incur this cost or benefit. Climate change, air pollution and most other environmental problems created by industries can be understood as environmental externalities wherein profits have been privatized but society as a whole must deal with the pollution created by these industrial processes. According to this logic (i.e. of protecting the environment by pricing environmental externalities) once nature's processes are given a financial value, prices of goods and services will reflect ecological costs and it will no longer make economic sense to produce ecologically harmful products or degrade environmental spaces. The project relies on the assumption that nature's processes can be safely disaggregated and effectively managed using natural capital accounting – often by substituting a destroyed ecological space with a preserved or regenerated space elsewhere. A whole new range of financial products will be created based on the concept of natural capital. This will generate lucrative opportunities for the financial sector with new markets for speculation. Expectations of profits will drive these new markets while environmental rhetoric provides convenient green marketing and public relations messaging to conceal continued – and even amplified unsustainable development.

The project is an amalgamation of conflicting agendas. On the one hand, it is a desperate attempt by scientists and environmentalists to convince industrialists and politicians to prioritize environmental concerns. On the other hand, it is the recognition by business of opportunities for profit in the creation of new green markets. Governments are already creating the infrastructure to sell the rights to the environment under the pretext that by conceptualizing nature as 'natural capital' corporate executives will be motivated to make responsible decisions.

Natural Capital and Ecosystem Services as Feedback Metaphors

The GEP has emerged from the development of the two foundational concepts: natural capital and ecosystem services.

Natural capital refers to the stock of natural resources that enable human existence. E.F. Schumacher coined the concept in his book *Small Is Beautiful* (2010, 2). The current meaning has its origins at the first Rio Earth Summit in 1992 (United Nations Conference on Environment and Development, UNCED) where it gained popularity in business circles. Now over four decades since the concept was first used, the idea has metamorphosed and the notion of nature as natural capital and as equivalent to capital in the bank is being adopted by some national governments.

Ecosystem services are the ecological processes than enable our lives. There are four types of ecosystem services:

1 **provisioning services** (producing food, fuel and fibre);
2 **regulation services** (the maintenance of the climate, regulation of floods and diseases, biological control of pest populations, pollination of crops and filtration and purification of water);
3 **cultural services** (benefits to society such as pharmaceutical products or spiritual, educational and recreational benefits); and
4 **supporting services** enabling all the above (nutrient cycling and the creation of soil, etc.).

While the concept can be a useful learning tool, the reduction of ecology into services that are helpful to humans instrumentalizes ecological relation and reduces the complexity of non-human nature. This becomes a problem especially when ecosystem services are used as a component of market processes as opposed to the context in which markets are enabled to exist.

These two concepts function as feedback metaphors. A feedback metaphor 'harbours social values [that] circulate back into society to bolster those very values' (Larson 2011, 22). These concepts emphasize an instrumental, utilitarian and fragmenting approach to the environment. As powerful metaphors, these ideas are being used to transform our relationship to non-human nature world by asserting that clean air, water, and habitats for endangered species are privileges that must be paid for – rather than a commons that all of humanity shares.

Economic Approaches to the Environment

Environmental economics, ecological economics, green economics and eco-socialist economics are distinct discourses with different conceptualizations of the relationship between the environment and the economy. (Figure 13.1). They have different policy prescriptions that relate especially to the degree to which they believe market mechanisms can help with the management of environmental problems. All too often economics is conflated with the ideological assumptions of neoliberalism, as if this is only type of economic policy is that is viable. Fortunately, there are alternatives. The theories of environmental economics (such as David Pearce) and, to a lesser extent, ecological economics (such as Robert Costanza)

	Liberal+ Neo-liberal	Environmental Economics	Ecological Economics type I	Ecological type 2 / Green Economics	Eco-socialist Economics
assumptions	value-neutral + composed of universal, unchanging laws	to bring the environment into economics	to bring economics into the enviornment	local, social, political and qualitative	capitalism exploits the environment and people
attitude towards planet	a source of scarce resources for the economy	a source for scarce resources and a sink for pollution	the system in which the economy is embedded *as a subsystem*	the context of human existance	the context of human existance
sees the environmental crisis as...	an inconsequential concern or an opportunity	a result of market failure	a result of regulatory failure	a result of the dismal and denial of nature	a result of capitalism's exploitative dynamics
principle concept	efficiency of markets	scarcity + efficiency of markets	steady state economics + a precautionary stance	degrowth + quality	commodity
policy impulse	market	market	regulation / market	participatory politics + decentralisation	participatory politics
belief in the concepts of 'externality' & 'substitution'	yes	yes	potentially	no	no

Fig. 13.1 *Economic Approaches to the Environment.* EcoLabs, 2017.

attempt to preserve biodiversity by factoring environmental externalities into economics, creating processes for valuation and trading of two particular externalities: environmental destruction and pollution. Ecological valuation processes value ecosystems (often in monetary terms) in order to use market-based mechanisms as a means of decision-making for the environment. While monetization processes are contested within the field of ecological economics, green economics and eco-socialist reject the monetization of nature entirely and oppose the use of the concepts of 'environmental externalities' and 'substitutability' (i.e. the assumption one ecosystem can be substituted for another). Since the UN's own data indicates that none of the world's top industrial sectors would be profitable if environmental costs were fully integrated (Roberts 2013), this analysis must be taken seriously. The chart in Figure 13.1 clarifies some of the key differences between various economic approaches to the environment.

Problems with the Natural Capital Approach

The UN's Green Economy project follows environmental economic theory closely related to a market liberalization agenda. Counter-movements claim that the project serves the interests of those who want continued unsustainable development without the interference of environmental regulations. It is based erroneous assumptions. The GEP re-imagines the global commons in a manner that demonstrates a misunderstanding of levels in ecological theory. Environmental economists hold that natural capital brings the environment onto the balance books of industry. But the environment is not only a supplier of resources but it is the geophysical context that makes the idea of resources even possible. This error is manifested throughout the new project. The global ecological commons are the source of life and the basis for all activities – economic and non-economic. Economics is a human construct made possible by ecological processes. Ecological processes are simply too

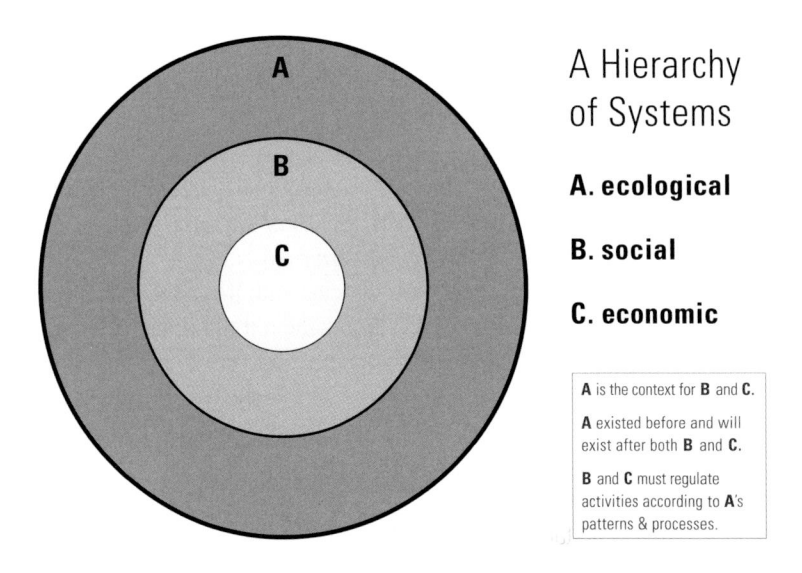

Fig. 13.2 *Conceptions of Human–Natural Relations: A Hierarchy of Systems.* EcoLabs, 2014.

complex to be captured absolutely through financial valuation processes because they are the *context* of economics, not a *subsystem* of economics (see Figure 13.2). Ecological systems are complex webs of interconnected relations that cannot be effectively understood, much less managed in isolation. Reducing the value of nature to financial terms facilitates exploitation in a material realm and these conceptions of the natural world as subject to the logic of the market are a prelude to the sale of those environments that are desired by industry for development. This conceptualization of the environment (i.e. in epistemological error) spawn multiple methodological and political problems.

Methodological Problems with the Financialization Agenda

Methodological problems on the financialization of nature include the limits of scientific capacities to value all of nature's processes; the consistent undervaluation of ecosystem services; the underestimation of risk; the issue of substitutability; and issues of communication and motivation. Humankind simply does not have the scientific capacity to measure all of the life-sustaining services provided by nature but what is obvious is that there will be no financial system to create wealth without the benefit of a stable climate, clean water and healthy local ecosystems.

When scientists do fix a price for nature, these values are often absurdly low. For example, the Prince's Foundation Accounting for Sustainability Project (2011) published an image (designed by Information is Beautiful) that inverts the hierarchy of the relationship between economy and ecology (see Figure 13.3). Here the global gross domestic product (GGDP) is illustrated as $63,000bn ($63 trillion) and the value provided by the Earth to the global economy is $50,800bn ($50 trillion) (Accounting for Sustainability 2012). This project aims to make a case for sustainability and yet the numbers misrepresent the relationship between the economic and ecological systems. GGDP would not exist without the Earth so it cannot have a higher value. With this way of conceptualizing the environment it appears as if the Earth is less valuable than even one year's GGDP. The numbers lend an aura of authority to environmental debates but the absurd calculations fail to establish a truthful framing of humankind's dependence on the natural world.

The natural capital agenda encourages the underestimation of risk. Profit-seeking market processes reward those who under-value ecological spaces and species. Ultimately the valuations have more to do with politics and power than the value of a particular ecosystem service. But even if the financial valuation processes were to give ecosystem services high value, the assumption that one ecosystem service can be substituted for another is inherently wrong. Ecosystem markets generate the conditions for the destruction of environmental spaces and species with the pretext that others will be conserved. While market processes give the 'impression that humankind can control nature as "assets" so as to have the possibility to "bail out" earth systems when they break down' (Fioramonti 2013, 118), once ecological thresholds are crossed, money cannot fix extinct species, collapsed ecosystems, climate change, etc. This project creates opportunities for business at first, but once destroyed an ecosystem cannot be saved by preserved ecosystems elsewhere.

Another set of problems pertains to communication, identity, values and psychological motivation. Motivational crowding out theory (Vatn 2000, 2010) describes how motivation for environmental conservation is impacted by utilitarian logic that risks 'eroding noneconomic

[Figure: Costing the Earth infographic showing $63,000 bn Global Gross Domestic Product, $50,800 bn Value provided by the Earth to the global economy, and $93 bn Needed to preserve the Earth's natural capital]

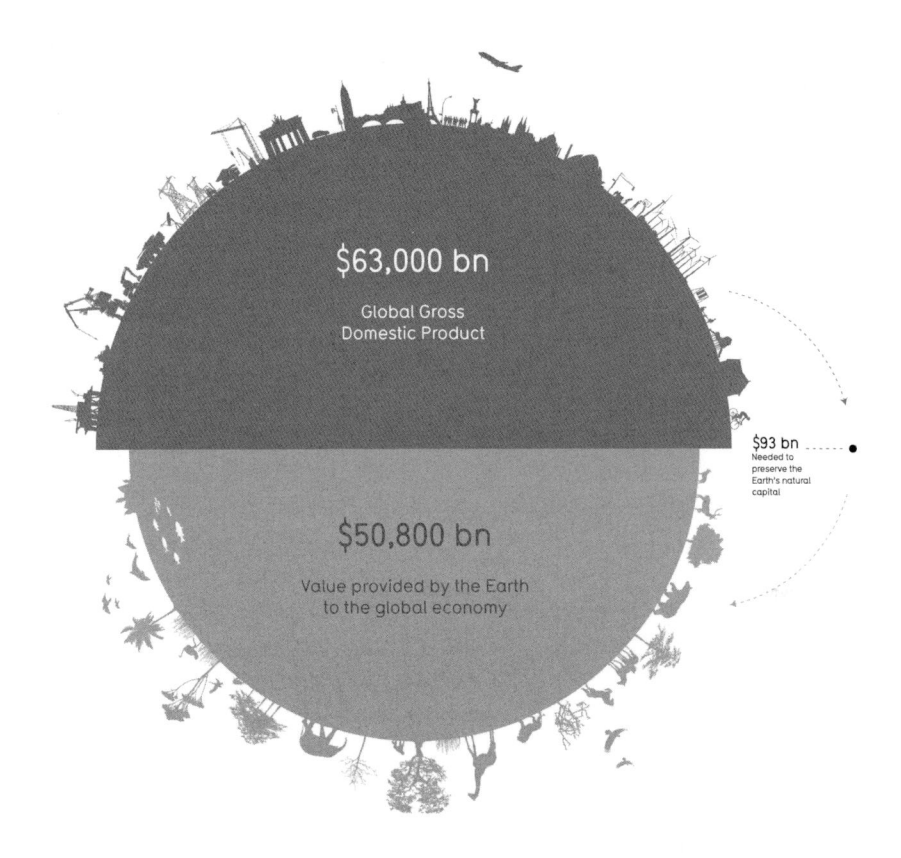

Fig. 13.3 *Costing the Earth* by Information is Beautiful Studio for The Prince's Accounting for Sustainability, 2011.

incentives for environmental stewardship' (Luck et al. 2012, 1024). The ways in which we frame human–nature relations matter (quoting George Lakoff):

> you cannot win an argument unless you expound your own values and reframe the issue around them. If you adopt the language and values of your opponents 'you lose because you are reinforcing their frame.' Costing nature tells us that it possesses no inherent value; that it is worthy of protection only when it performs services for us; that it is replaceable. You demoralise and alienate those who love the natural world while reinforcing the values of those who don't. (Monbiot 2014)

Environmentalists who accept the economic framing and monetization of nature are 'stepping into a trap their opponents have set' (Ibid.). Cognitive scientists have demonstrated the limitations of quantitative, utilitarian and exclusively rational modes of reasoning in motivation on politicized issues. The Common Cause project on motivation describes the ways in which human identity and values are encouraged or discouraged through social practices and communication. Their research suggests that financial valuation of nature encourages extrinsic values (values centred on external approval or rewards, values such as seeking social status) resulting in a simultaneous erosion of intrinsic values (values such as benevolence, care, empathy) (Crompton & Kasser 2009; Crompton 2010; Holmes et al. 2011,

Common Cause 2016). Furthermore, the utilitarian mindset established by quantification processes pushes out intrinsic values and the strong attachments to nature that have traditionally driven pro-environmental behaviour (Crompton 2013). Framing of the environment in monetary terms and as assets has profound implications that impact human motivation and agency.

Political Problems: The Neoliberalization of Environmental Policy

Political problems include the ruin of democratic participation in environmental decision-making associated with the neoliberalization of environmental policy. Evidence of the democratic failure of the project was evident at Rio+20 where NGOs, social movements and indigenous communities objected to the exclusion of their voices. With ecosystem services markets, democratic control of development agendas will be even more difficult as markets become the spaces where environmental decisions are made. Ecosystem services markets will become a new means of producing profit from activities that were previously not managed through commodity relationships.

As a mode of governance characterized by the elevation of market-based principles and techniques to State-endorsed norms (Peck 2013), neoliberalism presents itself as reigning in the power of the State. (One of its goals is to have public services provided by the market.) But under neoliberal governance what actually happens is that the State becomes larger it assumes new roles while dealing with the crises its processes create. The State shifts its focus from the provision of public services to funding of privatized services such as security, the military and prisons with subsidies for industries. Dramatic shifts in capital and power are masked behind rhetoric of austerity and security. The social fallout is colossal.

A growing body of literature on the neoliberalization of nature describes its characteristics: financialization, marketization, privatization, deregulation and reregulation (Castree 2008; Arsel & Buscher 2012; Sullivan 2013). Methods of governmentality are being implemented wherein complexity replaces responsibility and where governments outsource responsibilities for regulation (Castree 2008; Peck 2010). While rolling back responsibilities of the state, neoliberal governments simultaneously roll out other types of state functions, creating an 'explosion of "market conforming" regulatory incursions' (Peck 2010, 23) including huge bureaucracies. The neoliberal political project subordinates 'social and environmental considerations to macroeconomic policy imperatives ... [such that once] macroeconomic objectives are determined, every other policy target is chiseled in accordance' (Nadal 2012, 15). The UN claims that the green economy is politically neutral (UNEP 2011, 7) but new political reconfigurations are concealed behind the pretence of the inevitable expansion of capitalist logic, that is, markets as a means of organizing ever-increasing domains of human existence. The confusion resulting from the use of environmental language to describe a project of market expansion, privatization, neoliberal governance and even violent dispossession serves the interests of those who would not like these dynamics to be clear. Ultimately, the project relies on the private sector for investment and in exchange for capital investment ownership and control will be granted to private corporations and financial institutions. The financialisation of nature project has support in the corporate and financial sectors because it is seen as it is an expansion of the scope of market, an exceptional opportunity to create new financial instruments based on new types of capital and new markets for financial speculation.

The reconceptualization of the natural commons as natural capital has deep reaching implications. The man who first coined the idea of 'natural capital' had strong ideas in regards to the use of financial valuation approaches to protect nature. In the same book where the term was first published in 1973 E.F., Schumacher wrote:

> To press non-economic values into the framework of the economic Calculus ... it is a procedure by which the higher is reduced to the level of the lower and the priceless is given a price. It can therefore never serve to clarify the situation and lead to an enlightened decision. All it can do is lead to self-deception or the deception of others; for to undertake to measure the immeasurable is absurd and constitutes but an elaborate method of moving from preconceived notions to foregone conclusions ... The logical absurdity, however, is not the greatest fault of the undertaking: what is worse, and destructive of civilization, is the pretence that everything has a price or, in other words, that money is the highest of all values. (2010, 27)

As the concept of natural capital is transformed, the intellectual work of the environmental movement is used to advance a project that negates its ideological foundations. Despite the green rhetoric there is a symbiosis between this discourse and the anti-environmental climate contrarian discourse, since the lack of regulation enables corporate power grabs and weakens capacities in the public sphere to monitor and regulate polluting activities. In appropriating the most powerful environmental concepts, neutering these ideas of their content and then using this language to market political projects that support its own aims, the neoliberal political project masks its own dynamics while making it appear as if theirs is a green agenda. Revealing tensions and obfuscations on issues of the green economy is a substantial task for those of us who believe we will not protect the environment by creating the conditions for it to be sold.

Notes

1. That is, green economics and some versions of ecological economics.
2. That is, environmental economics and some ecological economics.

14 The Technofix

Conflicting ideas about the state of the environment and the role of technology in addressing its problems are at the crux of the sustainability predicament. The one extreme holds that design, technology and capitalism itself are all working well – although they all may need a little tweaking to adjust for climate destabilization (if climate change is even acknowledged). The critical position is that the politics of science and technology needs to be much more rigorously approached to avoid cascading environmental and social problems. The clash of perspectives over the existence and severity of environment problems is reflected in attitudes towards technology. Technology is often seen as politically neutral. However, the institutions that invest in particular technologies have political and economic priorities that are imprinted onto the social relations that their new technologies make possible. These priorities are reflected in what type of progress technological innovation supports. For this reason, progress itself is a contested idea and any political project that dismisses risks associated with accelerated technological progress is suspect.

Within this perplexing terrain, strategies and technologies evoked as pathways to sustainability must be approached critically. Relevant questions for interrogating the relative merit of particular technological claims include: What are the potential social and environmental consequences? Whose interests are being served? Whose interests are denied? Which priorities are served by the particular technology in question? Does a technology support justice in labour relations? Does it encourage asymmetric power relationships and thereby reinforce or even create new structural inequalities and/or authoritarian regimes? Questions such as these are particularly relevant with new technologies and economic strategies to address environmental problems. All too often technologies are developed without consideration of the social, economic and environmental justice issues at stake, often to the advantage of the very financial interests who are culpable of creating the initial conditions of ecological crises and grave social injustices.

Evidence such as the deteriorating state of oceans, air, forests, topsoil, upper atmosphere, aquifers, lakes, icecaps, mountaintops and biodiversity are clear indicators that we are (as a civilization) far from sustainable. The short-term interests of the vast majority of people on the planet and the long-term interests of all of humanity are not well served by the particular type of innovation strategies that are presently most aggressively promoted in this economic system. Despite decades of environmental campaigning and developments in appropriate and sustainable technologies, unsustainable development continues. Alongside sustainable options, disastrous products thrive, often in the homes of people who consider themselves to be conscientious consumers. Planned obsolescence is clearly still happening nearly six decades since Vance Packard first wrote about it in *The Waste Maker* (1960). A coffee machine that was invented twenty-five years ago creates completely unnecessary plastic waste with each cup of coffee. The progress sustainability pioneers are making in some spaces is dwarfed by aggregate pollution thanks to innovations such as the K-cup.[1]

Technological innovation is often presented as the solution to environmental problems. The technologies that assume this curative role are known as technofixes. The book *Techno-Fix: Why Technology Won't Save Us or the Environment* is endorsed by dozens of prominent environmentalists whose praise is listed in first few pages. Herman Daly claims: 'Salvation by technological advance and unlimited growth is the blind dogma of our day' (Huesemann & Huesemann 2011, iii). Norman Myers is more charitable: 'for the most part, techno-optimists have been simply misinformed and stand in urgent need so some extensive homework on this issue' (Ibid., iii). Technofixes are popular with governments, corporations and media because they serve the interests of powerful constituencies:

> Technofixes are very appealing. They appeal to who want huge projects to put their name to. They appeal to governments in short electoral leaders cycles who don't want to have to face hard choices of changing the direction of development from economic growth to social change. Technofixes appeal to corporations which expect to capture new markets with intellectual property rights and emissions trading. They appeal to advertising-led media obsessed with the next big thing, but too shallow to follow the science. They appeal to a rich-world population trained as consumers of hi-tech gadgets. They appeal to (carbon) accountants: technological emissions reductions are neatly quantifiable, if you write the sum properly. Technofixes appeal, in short, to the powerful, because they offer an opportunity to maintain power and privilege. (Fauset 2008, 7)

Those critical of technofixes claim that the unintended consequences of new technologies are often not known until it is too late; that technology within the current system does not address the most fundamental dynamics driving ecological harms; and that pinning our strategy to deal with environmental problems primarily on technology enables the dominant social and political dynamics creating ecological harms to go uncontested. One of the key contributions of social movements in this area is the mantra: 'there is no such thing as a technical solution for a social and political problem'. Technical approaches that aim resolve environmental problems without addressing the underlying causes of these problems are favoured by those who are content with a system that privileges overproduction, overconsumption and disproportionate benefit to a few.

The debate about technology involves tensions between those concerned with unintended consequences and those who deny these concerns. The socially positive potential of technology depends on social and institutional capacities to transform micro and macro-level barriers that prevent technology from fulfilling socially and environmentally beneficial functions. What is currently happening with potentially transformative technological strategies such as the circular economy and 3-D printing (discussed below) demonstrates how conventions and structural factors within the prevailing economic system sabotage sustainable transformations. Current ideological blindspots determine that the deployment of these strategies is not effectively addressing ecological crises on a viable scale. The barriers to effective implementation are simultaneously the personal assumptions of decision-makers putting these concepts into practice as well as system-level incentives that consistently thwart attempts to construct sustainable alternatives.

Environmental Scepticism and Anti-Environmentalism

Dominant discourses in the media, industry and academia celebrate technology while systematically dismissing its ecological and social consequences until problems are impossible to ignore. These ecoist discourses are the foundation of industrial society. Peter Jacque explains how environmental scepticism

> denies the reality and importance of (particularly global) ecological changes by specifically challenging environmental science. Skepticism refuses the assertion that there is a serious threat to human continuity or sustainability. This skepticism is essential for continued legitimacy of the dominant consumptive paradigm of industrial social order, particularly within the United States, because it deflects challenges to its assumptions and practices, including a brutally unsustainable hydrocarbon energy base.
>
> It should be clear at the outset, however, that this is not the same as classical skepticism, which refuse to decide on a conclusion until a maximum threshold of evidence is fulfilled. Rather, this is one front within a political movement that is skeptical of environmental protections and precautions like global warming policy, but very faithful to the dominant social paradigm of laissez-faire capitalism, technological solutions, growth, and Western materialism – indeed, to what Bill Hipwell calls 'industria.' Industria is a world-predatory system of knowledge and power found in the current global industrial network of corporations, states, military apparatus, and the 'elites who control them'. Environmental skepticism is a story from industria, consistent with what Val Plumwood calls the 'master story' or part of what Audre Lorde referred to as the 'master's house' which is always about domination and colonization. (2005, 435–436)

The anti-environmental discourse of 'industria' is the default position of those parts of corporate culture that do not even bother pretending to be sustainable (e.g. K-cups, Shell, Hummer, Monsanto) and their advocates. This discourse presents technology as neutral – and society as progressing towards greater prosperity and freedom. It ignores environmental concerns, increasing inequality and the emerging authoritarian nature of neoliberal governments.

Anti-environmental discourses are popular with a variety of constituencies. Associated campaigns, including climate denial (contrarian), are well funded (Brulle 2014). The resulting disinformation is widespread on television news and in the US government's federal Committee on Science, Space and Technology. The contrarian agenda is enabled by policy that disables regulation, enables corporate lobbying and facilitates corporate influence in policymaking. With carefully crafted images, design buttresses the 'master's house' that depends on the dismissal of the environmental and social impacts of technological innovations. Meanwhile, the environmental movement aims to expose environmental problems and the mechanisms of denial that enable these exploitative dynamics. Those concerned with the unintended consequences of technology (e.g. nuclear power, fracking, biofuels, agribusiness, geoengineering) fiercely contest the ecoist perspective that is still the dominant narrative in mainstream media and corporate cultures.

Growing Inequalities: An Economy for the Few

The environmental justice perspective argues that dramatic polarizations in wealth have never been higher, that poverty today is the result of inequitable distribution and that environmental problems impact the lives of the poor most severely. Dramatic increases in inequality are evident on a global scale:

- The richest 1 per cent of the world's population are getting wealthier, owning more than 50 per cent of global wealth. Oxfam and Credit Suisse both claim the 2015–2016 were the benchmark years that the combined wealth of the richest 1 per cent overtook that of the other 99 per cent of people (Credit Suisse 2016, 4; Oxfam 2017, 2).
- According to Credit Suisse' *Global Wealth Report 2016* half of all adults own less than $2,222 (30) while the top 10 per cent of global wealth holders have access to at least $71,600 and the top 1 per cent have over $744,000 (11).

Increasing inequalities and lowering standards of living for the poor in wealthy countries is widely documented (OECD 2011, 2015; Stiglitz 2013; Oxfam 2014, 2017; Piketty 2014; Reich 2015; Credit Suisse 2016). As globalization and automation result in mass underemployment, shifts in the distribution of wealth over recent decades have seen the evisceration of the working and lower middle classes.

The increasing inequality is most dramatic the United States which has seen a huge fall in living standards for the bottom half of its population. The past few decades have transformed the working class in America: 'Fifty years ago, when General Motors was the largest employer in America, the typical GM worker earned $35 an hour in today's dollars. By 2014, America's largest employer was Walmart, and the typical entry-level Walmart worker earned about $9 an hour' (Reich 2015). In the United States 'over the last 30 years the growth in the incomes of the bottom 50% has been zero, whereas incomes of the top 1% have grown 300%' (Piketty 2014; Cohen 2016; Oxfam 2017, 2). These trends have intensified since the 2008 economic crash during which time the wealthy have dramatically increased their share of total wealth. Joseph Stiglitz reports:

> From 2007–2010, median wealth – the wealth of those in the middle – fell by almost 40%, back to levels last seen in the early 1990s. All the wealth accumulation in this country has gone to the top. If the bottom shared equally in America's increase in wealth, its wealth over the past two decades would have gone up 75 percent. Newly released data also show that those at the bottom suffered even worse than those in the middle. Before the crisis, the average wealth of the bottom fourth was a negative $2,300. After the crisis, it had fallen sixfold, to a negative $12,800 ... [meanwhile] Adjusted for inflation, median household income in 2011 ... was $50,054, lower than it was in 1996 ($50,661).
> (2013, xii)

Specific government policies have resulted in an ever-rising concentration of wealth (Piketty 2014) – the transfer of wealth from the poor majority to an increasingly tiny and very rich minority. In the seminal book *The Spirit Level: Why Equality Is Better for Everyone* Richard Wilkinson and Kate Pickett demonstrate how inequality is harmful to society as a whole and to the economy. Despite the value of these insights, trend of increasing inequalities is likely to continue as long those with political power are blasé about the structural factors that perpetuate the polarization of wealth. It is not coincidental that the countries with the most dramatic growth in inequalities are those most fiercely committed to neoliberal economic policy.

Within the context of these dramatic demographic changes, environmental harms are concentrated on those people who cannot afford to avoid them. Environmental injustices are evident where industry is located near deprived neighbourhoods resulting in toxic water, air pollution and other environmental hazards. Inequality influences in the ways that society organizes the technological sphere. Power imbalances lead to development of particular types of technology that serve the interests of some people at the expense of others.

Post-environmentalists and Ecomodernists

Over the past few decades a new corporate-friendly environmental discourse has emerged that acknowledges environmental problems and offers an optimistic vision of solving these with technological innovations and certain types of government policy. The ecomodernist discourse assumes that environmental problems will be solved by decoupling growth from natural resource consumption and pollution, primarily through innovation in the technological and finance systems. The 'post-environmentalists' of the Breakthrough Institute (BTI) think-tank lead by Ted Nordhaus and Michael Shellenberger, exemplify this position. Their *An Ecomodernist Manifesto* (2015) calls for predominantly technocratic responses to environmental problems. The manifesto's authors (Michael Shellenberger, Ted Nordhaus, Stewart Brand, Mark Lynas, Roger Pielke Jr, and others) share a commitment to policy proposals that aim to address climate change without disturbing economic growth. They rely predominantly on nuclear energy for the cutting carbon emissions but have no apparent concern for the problems of nuclear waste or the fact that government subsidies are necessary to make nuclear a viable option.

The manifesto was spectacularly unpopular with environmental scholars. Bruno Latour was once a fellow at the Breakthrough Institute but broke with the group after the publication of this manifesto. During this period, he rebuked the ecomodernists:

> Never in history was there such a complete disconnect between the requirements of time and space, and the utopian uchronist vision coming from intellectuals. Wake up you ecomoderns, we are in the Anthropocene, not in the Holocene, nor are we to ever reside in the enchanted dream of futurism. (2015)

Other environmental scholars wrote scathing critiques of the manifesto: 'There is nothing really "eco" about ecomodernism, since its base assumptions violate everything we know about ecosystems, energy, population, and natural resources' (Caradonna et al. 2015, 16). Sam Bliss and Giorgos Kallis collected evidence that contests the ecomodernists' main claims (see Figure 14.1). The counter claims are:

1 urbanization *increases* land and resource use;
2 intensified agriculture *does not* spare land;
3 nuclear energy *cannot* fuel the world;
4 new technologies often *add* additional impacts to old ones rather than substituting them;
5 developed countries damage the environment *more*, not less;
6 material consumption *has not* peaked in wealthy countries; and
7 that quality and length of life *does not* depend on modernization (Bliss 2016; Bliss & Kallis 2017).

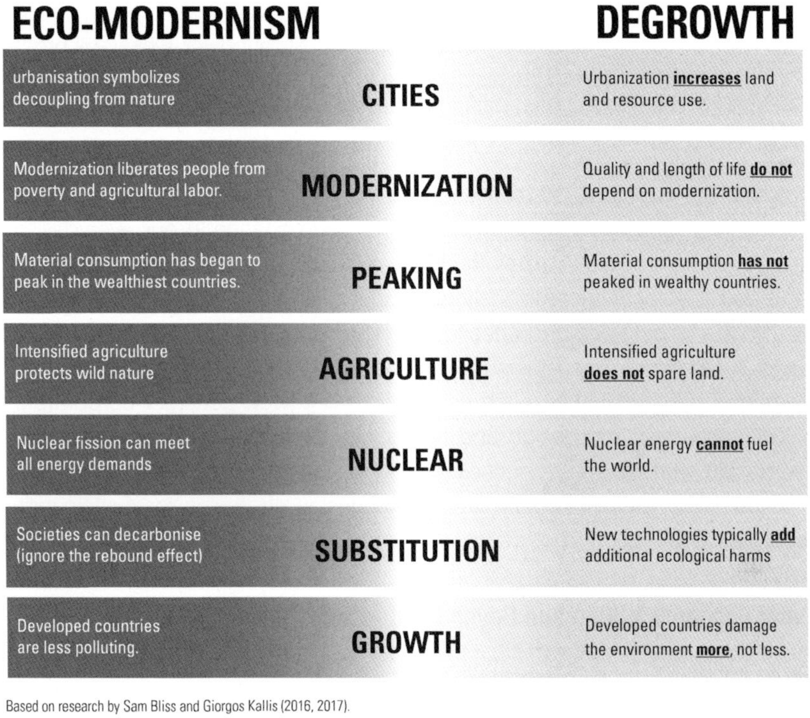

ECO-MODERNISM		DEGROWTH
urbanisation symbolizes decoupling from nature	**CITIES**	Urbanization **increases** land and resource use.
Modernization liberates people from poverty and agricultural labor.	**MODERNIZATION**	Quality and length of life **do not** depend on modernization.
Material consumption has began to peak in the wealthiest countries.	**PEAKING**	Material consumption **has not** peaked in wealthy countries.
Intensified agriculture protects wild nature	**AGRICULTURE**	Intensified agriculture **does not** spare land.
Nuclear fission can meet all energy demands	**NUCLEAR**	Nuclear energy **cannot** fuel the world.
Societies can decarbonise (ignore the rebound effect)	**SUBSTITUTION**	New technologies typically **add** additional ecological harms
Developed countries are less polluting.	**GROWTH**	Developed countries damage the environment **more**, not less.

Based on research by Sam Bliss and Giorgos Kallis (2016, 2017).

Fig. 14.1 *Degrowth vs Ecomodernism.* EcoLabs, 2016.

Despite evidence put forward by the rest of the environmental movement, on the defensive and the offensive the ecomodernists spend more time attacking environmentalists than challenging the power brokers who propel environmental problems (fossil fuel interests, etc.). Playing the role of environmentalists who devote significant time and resources into attempting to dismiss and denigrate the environmental movement and who do not critique industry or carbon intensive development policy is a strategically lucrative position for the ecomodernists. They are thereby attractive partners for 'industria'. Giorgos Kallis warns: 'eco-modernization is an oxymoron' (2015).

Apolitical Design: Default to the Unsustainable

Deeply entrenched in the construction of consumer capitalism, designers may be tempted to accept the ecomodernist discourse that is so easily absorbed into the current paradigm. To some, ecomodernism appears rational, neutral and professional. To those concerned and knowledgeable about the ecological and social consequences of various current and proposed technologies and industries, what this discourse reveals is an absence of care. An absence of care is often masked by evading the political (sometimes by people who claim to care a great deal). Design critic Rick Poynor describes the consequences of designers' avoidance of political realities:

> The problem that dogs all analysis of these issues as they relate to design is that, at root, they are political questions. Any discussion that fails to acknowledge this will never

move very far beyond vague, platitudinous statements of a desire to 'improve things' ... if you do not acknowledge the reality of struggle for political power, which will proceed regardless, you simply play into the hands of those who have no compunction in exploiting your lack of 'voice'. So what design badly needs is people ready to stand up and speak out. (2006, 60)

Disengagement is the default position enabling the status quo and perpetuating unsustainable conditions. Sustainability depends on disruptive change in our everyday lives, in corporate culture and in politics. It relies on engaged change makers with critical perspectives on power. It also relies on a rigorous analysis of the environmental and social consequences of various models of progress and specific technologies. It depends on work to dislodge structures that reproduce unsustainable conditions. The disengagement from concerns of environmental and social justice is a capitulation to power: generally corporate power. A critical orientation towards design and technology is necessary to assess the viability of specific design strategies and technological innovations in terms of their contributions towards sustainability. Below I briefly review two design strategies for sustainability: the circular economy and 3-D printing.

The Circular Economy: Within and Beyond 'Industria'

Strategies for the design for a circular economy promise to create no waste systems wherein all material will be reused. Like other potentially transformative design practices, the concept will be valuable to the extent that ecological values and principles are integrated on all scales. Circular economy projects have made impressive progress but this work continues to be undermined by sets of assumptions and system structures.

The cradle-to-cradle design concept was coined and developed by Walter R. Stahel in the 1970s. The circular economy model went mainstream with the book *Cradle to Cradle: Remaking the Way We Make Things* (2002) by co-authors William McDonough and Michael Braungart. The pair's ecoeffectiveness approach breaks all products into two spheres, material flows and waste steams: (1) the biosphere (the biological and biodegrabable) and (2) the technosphere (technical nutrients and industrial systems) (McDonough & Braungart 2002, 92). The material in the first sphere can be composted. These material flows must be separate from the second sphere which is endlessly recycled: 'To eliminate the concept of waste means to design things – products, packaging and systems – from the very beginning on the understanding that waste does not exist' (Ibid., 104). Toxins must not enter the organic metabolic process but can be more handled safely in the technical metabolic sphere.

The framework is a valuable contribution to sustainable design and the authors became environmental celebrities (McDonough was hailed in *Time* magazine's 'Heroes for the Planet'). But behind the scenes everything was not so impressive. A 2008 article *Fast Company* described how Macdonough's propitiatory approach sabotaged his projects and the cradle-to-cradle method. McDonough alienated partners with litigious behaviour and attempted to 'make a royalty off of every green standard and every green product out there' (Sacks 2008). The problems with the circular economy approach were evident not just in McDonough's behaviour but in the ways the idea was used by institutions.

In the UK, attempts to launch the circular economy have been made significant progress over the past decade – but there have also been hiccups. The circular economy requires *systems and processes* that dramatically reduce waste, but all too often organizations limit the scope of enquiry to profit-making products. Designers must break out of habits that locked-in linear economics to avoid reproducing current production assumptions the perpetuate cradle to grave design.

Circular economies have not yet been able to achieve their goals on scale because the social norms and political structures of industria are barriers to sustainability. Strategies such as the circular economy are thwarted by common assumptions including:

- Sustainability can be achieved at individual product level rather than the system level.
- Profits rather than social and ecological benefits can remain the primary priority in sustainable design.
- Privatizing profit and socializing ecological harms is compatible with sustainability.

No amount of circular design concepts infiltrating corporate culture will effectively address sustainability challenges as long as these initiatives ignore the structural dynamics of the context in which they operate. The 'inability to recognize the very difference between consumption to meet basic needs, commercial consumption, and excessive consumption only sustains the power relations that exacerbate the overconsumption' (Parr 2009, 155). The circular economy is a good idea with the potential to transform industrial practice and dramatically reduce waste – but it will not do so until we transform the assumption and structures that keep good ideas from addressing problems on scale.

3-D Printing: Beyond Pointless Plastic Products

Despite the hype, 3-D printing environmental consequences current eclipse its potential for socially valuable work. Rachel Amstrong explains 'urgent rethinking is required to avoid the revolutionary potential of 3-D printing being lost in a sea of pointless plastic products' (2014). As is the case with other technologies, full life cycle assessments (LCAs) and wider environmental and sociological evaluation of how this technology is deployed are required to evaluate impacts. Armstrong claims:

> if 3-D printing is genuinely going to take us beyond yet another version of 'less damaging' industrial practices – that are cheaper, faster, better – into a space where working with these systems could shift our values and, for example, produce objects that are genuinely life-promoting, then it needs to respond to this emerging cultural concern and practically engage in establishing a practice of 21st-century materialism. (2014)

Beneficial outcomes will depend on the sustainability knowledge and priorities of those who use this technology and those who create policy on its deployment. Currently, 3-D printing typically reproduces problems with more accessible and often more wasteful manufacturing processes.

There are benefits of 3-D printing: environment and otherwise. Environmental benefits are achieved due to the fact that 3-D printing is an additive process (it is also known as additive manufacturing). This means there is potentially less waste in the production process (unlike subtractive manufacturing where products are cut from a block of material). Products can be printed when and where they are needed. Small complicated items can be produced (such as solo parts to fix broken appliances). Decentralized DIY maker cultures are emerging that challenge traditional manufacturing in huge factories overseas. Other benefits include accelerating new product development and the ability to offer customized and limited-run products such as new prosthetic limbs.

Nevertheless, the unintended consequences of 3-D printing are significant. The 3-D printers are energy intensive – about 50–100 times more electricity than injection moulding (Kurman & Lipson 2013) and hundreds of times more than traditional casting where metal powder is fused together (Gilpin 2014). They typically rely on plastics (printers usually use either fossil-based ABS thermoplastics or plant-derived polylactic acid, PLA) that end up in landfill. They produce unhealthy air emissions. The embodied energy in the machine itself is very high. Creation is slow and labour intensive. This is a technology that can only be considered sustainable when it uses biodegradable material and renewable energy at the point of production and when the embodied energy of the machine itself is accounted for. For this last reason, sustainability will be considerably enhanced if production is community-based rather than decentralized into individual studios or homes. In this way the embodied energy in one machine is shared between many users. There is 'considerable scope for improvement in the environmental impacts of 3-D printing. The starting point can be a proactive consideration of environmental factors from the outset of production/produce design' (McAlister & Wood 2014, 213). Considering the potential environmental harms of this technology in the context of throwaway cultures used to extravagant waste, 3-D printing is more often than not simply a means to create vanity projects and exacerbated environmental harms. While there are examples of sustainable 3-D products and architecture, the last thing this planet needs is more plastic trickets.

The Luddites and the Roots of Commonality

All technologies should be scrutinized for their environmental and social impacts. Those who would like to focus on the unintended consequence of new technologies are often characterized dismissively as Luddites. The term 'Luddite' has its origins in the Luddite Rebellion (1811–1813) after Ned Ludd (later known as King or General Ludd). The Luddites are one of the most poorly understood historical social movements. The Luddites were skilled textile artisans and workers in Nottinghamshire, Yorkshire and Lancashire. They were the high-tech workers of their day: the craft labourers who were the most familiar with the machinery. The Luddites resisted and destroyed a *particular type of textile machine* in response to the ways that this *specific* new technology destroyed their livelihoods and communities. These machines were described in a Luddite text as 'hurtful to commonality' (1812 quoted in Binfield 2004, 209). 'Commonality' means the common good, in the tradition of the commons (that are managed communally by locally people). The resistance of the Luddites wasn't about *all technology* but about machines being used to take away jobs and establish profoundly unequal power relations. The Luddite Rebellion was brutally crushed by the State.

The modern usage of the term 'Luddite' originates from the 1950s when CP Snow and other technocrats pushing nuclear power and a modernization agenda in Britain used it to characterize their opponents. Today, the Luddites are significant because: 'the destruction of the Luddites by the State established the principle that industrialists have the right to continually impose new technology, without any process of negotiation, either with the people who have to operate it or with society at large' (King 2015b). Activists concerned with the politics of technology use the concept of technocracy to refer to a culture of modernity in which technical discourses have hegemonic status above all other ways of thinking. Thus for critical technology campaigner David King, the problem is not technology but technocracy (2015a, 2015b). 'Technocratic' refers to a situation where decisions are made by technical workers without social and political considerations. Technical workers are assigned their decision-making role by the established regime of power. Technical decisions conceal values embedded into institutional practice: usually maintaining and reinforcing current power structures with a technofix. These decisions are made without democratic debate regarding potential unintended social and ecological consequences.

Beyond Technocracy and Ecomodernism

Without a doubt technologies have created amazing conveniences, longer lives, pleasurable activities and extravagant wealth for many people. Today renewal energy, biomimicry and other appropriate technologies hold potential to catalyst dramatic techno-sociological transformation that could contribute to regenerative ways of living on this planet. Unfortunately, the aggregated impacts of technological progress are also causing climate change and other severe environmental problems and injustices. Specific technologies are hurtful to commonality. Almost all sustainability advocates agree that various types of technology must be activated for ecologically viable ways of living to become possible. With the recognition that social and environmental problems are commonly shared (although often unevenly distributed and impacting the poor most severely) decisions about technology and environmental conservation must be political decisions. Critical thinking about the social and environmental impacts of particular technologies is not about closing down options but about being truthful and acknowledging that technology has implications. These sometimes destroy lives and environments far removed from those making the decisions about which technologies are chosen and implemented.

Note

1. Co-inventer of the K-Cup John Sylvan sold his company (Keurig) in 1997 and is now part of the anti-K-cups movement #KillTheKCups. Unsustainable design has a legacy beyond the intentions of its creators.

15 Data/Knowledge Visualization

Data visualization makes big data and other information accessible and meaningful in ways that reflect both the explicit intentions and the implicit assumptions of designers. Despite efforts some designers make to be neutral and objective interpreters, all information design is embedded with suppositions. When data visualization illustrates trends and presents truth claims it privileges certain perspectives. We all rely on accurate information that effectively captures the complexity of contemporary conditions but neither data itself nor data visualizations are politically neutral. Data reflects power relations, special interests and ideologies in terms of which data is collected, what data is used and how it is used. In this chapter I take a critically informed approach to data and knowledge visualization. Due to the inherent reductionism in data visualization, it can easily be used in ways that obscure complex phenomenon. For this reason, in many instances, knowledge visualization is a more effective and honest approach. This is especially true on issues of controversy.

Data Visualization Does Political Things

In order to understand how meaning is constructed with data visualization and what can go wrong, it helps to theorize the various practices involved. All visualizations are interpretations that can only capture partial information. Data visualization is created in a process involving multiple decisions. The decisions involving which data to collect, which data to illustrate, how to illustrate it and where to disseminate it are decisions that reflect assumptions, unstated (often unacknowledged) ideological perspectives and subjective judgements. With these choices, some information will always be missing.

In an article called 'What Would Feminist Data Visualisation Look Like?' Catherine D'Ignazio uses feminist standpoint theory to analyse the ways that data visualization works to communicate information. Standpoint theory holds that 'knowledge is socially situated and that the perspectives of oppressed groups including women, minorities and others are systematically excluded from "general" knowledge' (D'Ignazio 2015). Data visualization often functions to make a particular point of view appear as the reliable position: 'Even when we rationally know that data visualizations do not represent "the whole world", we forget that fact and accept charts as facts because they are generalized, scientific and seem to present an expert, neutral point of view' (Ibid.). Yet data visualization reveals certain phenomenon while simultaneously concealing more complicated realities. In a world with dramatic power imbalances, some people's interests are represented at the expense of others. For this reason D'Ignazio claims that data visualization must be understood as 'one more powerful and flawed tool of oppression' (2015, 2). When it amplifies the point of view of the powerful, excludes the perspectives of the disempowered, over-simplifies complex and controversial issues and then represents the results as an authoritative view data visualization does profoundly political things.

With all types of information design meaning is constructed by both the designer and the audience. Cultural theorist and sociologist Stuart Hall describes how communication is encoded by a producer and then decoded by an audience in ways that are processes of cultural translation (1973). In both instances, interpretations occur. Designers construct data visualizations by selecting relevant data sets and organizing information to tell a particular story in order to reveal certain elements about a situation. Audiences decode visualizations by interpreting visual strategies and codes according to cultural conventions and their own assumptions. At the production and the reception stage of communication processes, meanings can be distorted. Approaching data visualization critically involves recognizing which stories are being told and what stories are not told about the same situation. Why are some perspectives communicated while others are not? How do narrative, style and the media itself reflect on the content? These questions are a starting point for designers and audiences concerned understanding a topic under investigation and the problems that can emerge in visualization processes. Data visualization can project an unwarranted veneer of objectivity that does not always warrant the authority that it assumes.

The Limits of Quantitative Displays

Quantitative comparisons and analysis is useful for some tasks and less useful for others. In Chapter 4 (p. 60) I describe how the dominance of quantitative methods has dramatic consequences in the manner that society is organized. Clearly, quantitative methods are a necessary foundation for analysis in many instances – but they are not necessarily the most relevant modes of analysis for many complex and controversial problems. Management theorist Margaret Wheatley explains that by condensing observable facts to numbers, the 'act of measurement loses more information than it gains' (2006, 65). This loss of information occurs as a result of at least two reasons:

1 In the process of reducing phenomena to numbers, qualitative information is discounted. Numerical data is an abstract type of information captured with quantitative values. Identified elements are counted with others of the same type in order to facilitate quantitative comparisons. The process of reducing phenomenon to a numerical value is a process that loses potentially important information that will then be missing from the dataset.

2 Decisions on how and what to count in the first place are decisions that reflect preconceived ideas and often the interests of those who collect the data (or sponsor the collection of data). The process of collecting and then visualizing data is never entirely without bias and never complete. Something will always be lost. The decisions of which data to collect and which data to use reflect norms and agendas that are then reflected in the datasets.

For these reasons, quantitative and numerical displays on their own are often reductionist and over-simplistic. Purely quantitative approaches fail to capture power relations, ideologies, attitudes, motivations and behaviours that cannot be reduced to a number. Quantitative methods are especially inadequate on politicized issues such as the environment. Data visualization embody values, reflected in which data is selected, as well as the methods, media and styles used to communicate information.

The Dangers of Digital Positivism

Digital positivism refers to the ways that digital methods, including data visualization, can assume an authority associated with empirical work and the hard sciences that masks subjective decisions, perspectives and ideologies. Media theorist Vincent Mosco describes digital positivism as digital work that prioritizes 'quantitative over qualitative data, arguing that the former provides the best opportunity for meaningful generalizations and that, when necessary, qualitative states can be rendered qualitatively' (2014, 196). Digital positivism advances a false certainty and a crude understanding of how data mediates knowledge. The danger here is not only that complexity is reduced to numbers, but that certain types of knowledge are prioritized as the expense of others: 'It is uncertain what is worse: that big data treats problems through over-simplification or that it ignores those that require a careful treatment of subjectivity, including lengthy observation, depth interviews, and appreciation for the social production of meaning' (Ibid., 198).

Digital positivist obscures value judgements that are embedded in data visualization. The resulting communication can be deceptive and a dangerous basis for decision-making. In a *New York Times* article called 'What Data Can't Do' David Brookes explains how data embodies hidden agendas:

> data is never raw; it's always structured according to somebody's predispositions and values. The end result looks disinterested, but, in reality, there are value choices all the way through, from construction to interpretation … This is not to argue that big data isn't a great tool. It's just that, like any tool, it's good at some things and not at others. (2013, 14)

Digital positivism reflects and fosters a false sense of epistemic certainty that enables risk taking on issues of the environment, technology and science. It is the presentation of data as more than a particular *interpretation* that is misleading. The over-simplistic presentation of data without reference to context or associated controversies can be irresponsible and unethical on issues of justice – as the perspectives of less empowered constituencies are typically ignored.

Digital positivism is evident where data is presented as it is the conclusive evidence. The *Our World in Data* project by economist Max Roser offers overviews of global trends with data visualizations. The project website claims it is 'visualising the empirical evidence' and the headline over charts reads: 'empirical view' thereby emphasizing the authority of the data. Problems with this approach are evident where the data visualizations obscure politicized topics. For example, on 10 May 2015 Roser (@MaxCRoser) tweeted 'Declining Racial Violence in the US since 1882' (2015a) and embedded a chart in the tweet displaying decreasing numbers of lynching in the United States over the past century (Figure 15.1). With this chart Roser picked a particular data set and used it to make a claim about racial violence in America (Roser 2015b). By visualizing antiquated data while ignoring more relevant data (such as police killings of black people) his chart bolsters a political perspective

Fig. 15.1 Tweet and chart: *Lynchings in the United States, 1882–1969.* Max Roser, 2016. Note the 'Empirical View' headline.

and serves a particular narrative. Meanwhile Black Lives Matter activists highlight police violence and have demonstrated that racial killings have been systematically dismissed in the American justice system. The FBI has not historically collected complete information on police killings[1] and so these statistics are only collected by newspapers and activists organizations (Swaine & Laughland 2015). With Roser's tweet on racial violence, decisions with political implications have been made at all stages: the necessary data was not counted by official bodies, the choice to use the lynching data to make charts, the choice to label and describe the chart as representative of racial violence, etc. These omissions are obscured by the presentation of this data as representative of racial violence in America.

The racial violence claim is a relatively easy example to critique since work has been done to gather missing data on police killings and activists have politicized this issue. Representations of the environment are also politicized, but typically it is much harder to identify exactly how these hyper-complex issues are obscured in communication processes. For example, Figure 15.2 'Global death rate from natural catastrophes (1900–2013)' supposedly

Empirical View

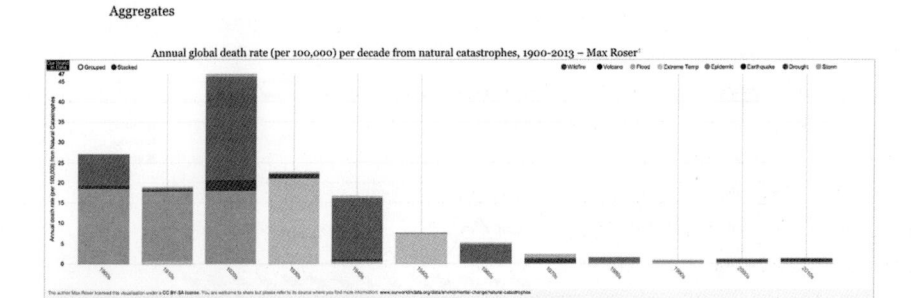

Fig. 15.2 *Natural catastrophes: Annual global death rate (per 100,000) per decade from natural catastrophes*, 1900–1913. Max Roser.

illustrates significant improvements in humankind's capacity to survive natural catastrophes (Roser 2015c). But the data reflects an approach that is very narrow in its boundary conditions. Considering the ways that deteriorated environmental circumstances trigger conflict and the current refugee crisis – how survival is defined and measured in the wake of natural disasters must also be a contested issue. While the data presented here could suggest that there are more survivors of natural catastrophes per capita in the short term, more extreme weather events are happening more often. The impact of climate de-stabilization is already dramatic in many places. The damage to livelihoods by extreme weather is a point of contention at yearly United Nations climate conferences. Nations on the front line of climate impacts fiercely contest the narrative that natural catastrophes are less of a problem now than in the past as whole islands disappear under the sea and millions leave regions struck with drought.

Obscuring Datawash, Missing Dark Data

Datawash is data visualization that conceals or obscures. Datawash over-simplifies, cheery picks and de-contextualizes information and presents claims that obfuscate complex information (Boehnert 2015, 2016). Datawash often serves an ideological function in promoting a particular agenda while it appears to deliver only the facts. Datawash is most evident on issues of controversy where corporate interests differ from public interests – such as is often the case with the environment. The associated concept of darkdata is the missing data (Corby 2015). It refers to what isn't measured or tracked. This missing data is often as important (or more) than the data that is collected. Where certain data is not collected this is often due to the epistemic and ideological assumptions of powerful constituencies – or simply where the availability of certain data is blatantly against their interests. For example, in the case of the chart above (Figure 15.1), the darkdata is the data set on police killings of people of colour. In cases involving the environment, darkdata can be associated with 'unintended' consequences as the implications of new technologies and development are not investigated and thus data on risks is missing.[2] This data is either unknown, difficult to collect or not collected for political reasons (and because of people's assumptions about what data is worth collecting).

The concepts of datawash and darkdata open discursive space by focusing attention on what is obscured, neglected and unknown. Roser presents his work as 'the empirical view' (see Figures 15.1 and 15.2) but his favourite data sets are by no means the authoritative data sets. The data sets he visualizes are those that suit his particular ideological perspective. It is worth repeating: data is not neutral. It is instead an assemblage of infrastructures, laws, social discourses, technologies and politics (Corby 2015). It reflects the practices, concerns and interpretations of social science research and the institutions that collect data. Data is 'informed by history, culture, and society, and these help define, reproduce, and shape the use of this data' (Räsänen & Nyce 2013, 659). Values, conventions, sensibilities and ideologies impact choices regarding which data is collected, as well as the methods, media and styles used to communicate data. Critical approaches to data visualization make these agendas explicit. D'Ignazio describes the characteristics of feminist data visualization as: (1) representing uncertainty, missing data and the provenance of data; (2) referencing the material economy behind the data; and (3) making dissent possible (2015). D'Ignazio's analysis is an example of how feminist theory informs critical positions that support analysis and good practice on a variety of politicized issues including environmental ones.

Enabling Knowledge Visualization

We live in an information rich world but information alone does not necessarily lead to understanding or the capacity to act in effective ways. For example, we have an abundance of data on climate change but we have not created effective means to adapt much less mitigate impacts of a de-stabilized climate system. Data visualization on climate change and other complex problems does not necessarily support effective action or policies to lower greenhouse gas emissions. As we have seen, sometimes data-driven methods do more to obscure the problem than facilitate solutions. On the other hand, communication design can help audiences move from processing information to deeper, more reflective learning. Knowledge visualization aims to create actionable understanding of new information by encouraging these deeper learning processes.

Knowledge visualization makes information actionable by putting data in context. Data becomes more useful, meaningful and less likely to work in obscuring ways once it is contextualized. Information theory clarifies the ways that communication and learning occurs at different levels:

Data is the pure and simple facts without any particular structure … the basic atoms of information,
Information is structured data, which adds meaning to the data and gives it context and significance,
Knowledge is the ability to use information strategically to achieve one's objectives, and
Wisdom is the capacity to choose objectives consistent with one's values and within a larger social context. (Logan & Stokes 2004, 38–39 quoted in Logan 2014, 35)

These four categories are illustrated on the *Wisdom/Knowledge/Information/Data Visualization Triangle* (Figure 15.3). Data is the most reductive type of information. Data is everywhere but in focusing on any particular dataset, one's perspective is reduced to one type of information, often captured with numerical data. As we have seen, while all

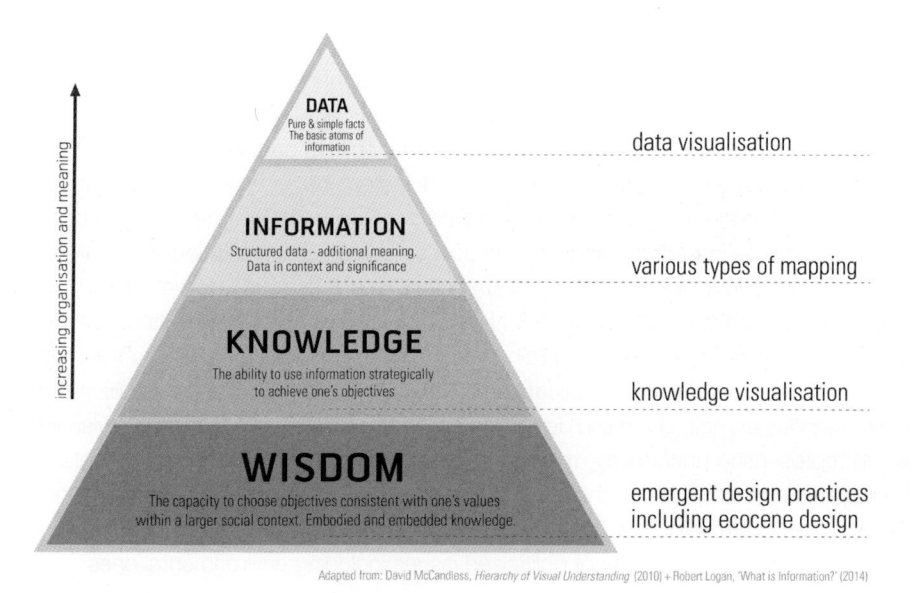

Fig. 15.3 *Wisdom/Knowledge/Information/Data Triangle.* **Adapted from David McCandless'**, *A Hierarchy of Visual Understanding?* (2010). EcoLabs, 2017.

visualization techniques reflect ideas of the people involved in production, data visualization is a practice where this partial view is easily obscured.

Knowledge visualization is a more expansive approach that displays multiple interpretations on different scales and from different disciplinary perspectives. Knowledge visualization is 'experiential and actuative' where getting someone to take action is the focus of the practice (Masud et al. 2010, 447). Learning theory tells us that deeper types of learning that are necessary for action on new information (especially on difficult topics such as the environment). Qualitative, interpretative and critical visual approaches present fuller sets of relationships that help audiences align new information with their own values, enable more integrated learning and the development of new agencies and capacities to take action. Knowledge visualization offers more meaningful approaches for the visualization of complex topics but it also presents significant challenges for designers in terms of the depth understanding necessary to develop these broad overviews.

Controversy Mapping

Mapping controversies is often a good starting point for learning about environmental problems and other politicized issues. The *Climaps by EMAPS: A Global Issue Atlas of Climate Change Adaptation* project was a three-year collaborative project funded by the seventh Framework Programme of the European Union. It is the largest yet experiment with the method of controversy mapping. The research group describes the controversy mapping method as:

> a research technique developed in the field of Sciences and Technology Studies (STS) to deal with the growing intricacy of socio-technological debates. Instead of mourning such complexity, it aims to equip engaged citizens to navigate through expert

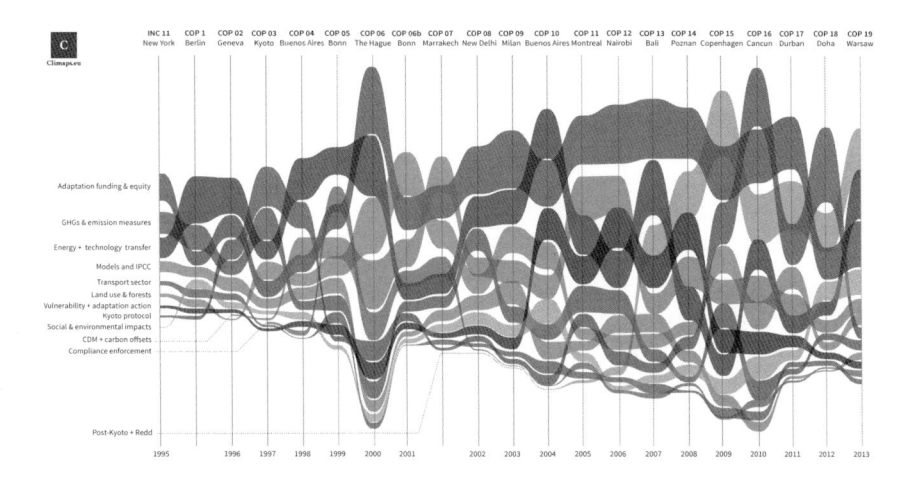

Fig. 15.4 *Rise and Fall of Issues in UNFCCC Negotiations, 1995–2013.* Climaps by Emaps, 2014.

disagreement. Instead of lamenting the fragmentation of society, it aims to facilitate the emergence of more heterogeneous discussion forums. (Venturini et al. 2014, 1)

The project features thirty-three issue-maps on climate change and the annual United Nations Framework Convention on Climate Change (UNFCCC) Conference of all Parties (COP) negotiations (see Figure 15.4). The designers addressed a 'classic simplicity/complexity trade-off: how to respect the richness of controversies without designing maps too complicated to be useful?' (Venturini et al. 2015). The maps in this series effectively capture an overview of debates, climate negotiations and policy emerging from the UNFCCC discussions over more than a decade. Figure 15.4 illustrates topics under discussion at the annual UNFCCC annual conferences.

What is not evident here in the series are the intense conflicts including outright contradictions in this highly contentious area. Within climate communication the same language is often used when referring to very different sets of interests and policy proposals. The meaning of language on politicized issues such as climate change differs according to its context. Misrepresentations in this area have serious consequences as policy is developed based on perspectives, interpretations and ideologies. Currently policy on climate change enables industrial systems that are undermining the stability of the climate system. Since ideologies, power relationships and contradictions are difficult to identify in most available datasets, qualitative and interpretive approaches are necessary when approaching issues of controversy such as climate change. Different discourses on climate reflect different ways of understanding political problems and different types of proposed solutions. In the highly politicized context of climate communication, these distinctions matter because they represent vastly different proposals for action.

Discourse Mapping

Discourses are shared ways of understanding the world that provide the basic terms for analysis and define what is understood as common sense and legitimate knowledge

(Dryzek 2013, 9). Diverse values, vested interests, critical perspectives and insights are embedded within discourses. These both reflect and construct attitudes towards the natural world. Discourse mapping is an interpretative method that reveals political agendas, diverging worldviews and ideological assumptions. It reveals the ideas that underlie various approaches to environmental problems. In displaying the relationship between discourses, the outlines of controversies are illustrated. Discourse mapping can highlight assumptions that are obscured by epistemological and ideological blind spots inherent within more reductive visualization practices. Discourse mapping captures meaning lost with data visualization.

The *Mapping Climate Communication*[3] project uses a discourse mapping method to present an overview of how climate change is communicated in the public realm by visualizing and contextualizing actors, events, actions within five dominant climate discourses: climate science, climate justice, climate contrarian, neoliberalism and ecological modernization. These five discourse are associated with five ways of framing climate change (as illustrated in Figure 15.6). The *Climate Timeline* (Figure 15.5) visualizes the historical processes and events that have led to the growth of various ways of communicating climate change. The *Network of Actors* (Figure 15.7) illustrates relationships between institutions, organizations and individuals participating in climate communication in Canada, United States and the UK according to the five discourses. In this project climate communication refers to all of the ways in which public understanding of climate change is developed through social communication processes. Since communication happens at the level of rhetoric as well as the level of action, discourses here include explicit messages and also messages that are implicit within political, corporate and non-profit organizational activities and policy. In other words, it includes communication by omission, that is, what is communicated by the denial or ignoring of climate change in places where it is relevant. This approach reveals tensions in climate communication by exposing contradictions between what was said and what was done about climate change. For example, institutional actors claim that climate change is a serious threat but continue to support carbon intensive development. Following debates over four decades it is clear that rhetoric often masks a lack of action. Since it is easier to say that low emissions are necessary than to do the political work that will make this possible, this conflict between explicit and implicit messaging is important. The discursive confusion that results from contradictory communication (evident in the neoliberal discourse) is theorized as central to the ongoing slow progress in climate policy.

The project explores communication problems that contribute to the continued rise in emissions despite the significant work by the climate science community and the environmental movement over four decades. All three climate discourses that acknowledge the

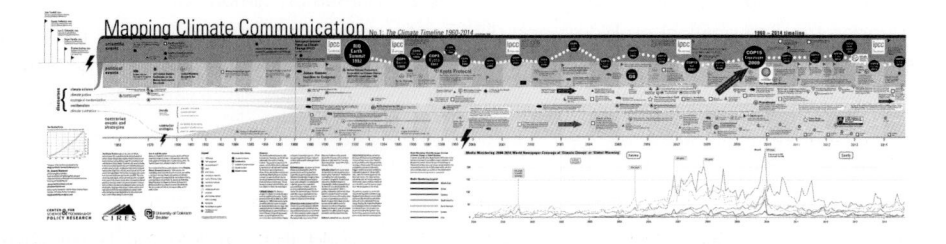

Fig. 15.5 *Mapping Climate Communication, No. 1 The Climate Timeline,* Boehnert, 2014.

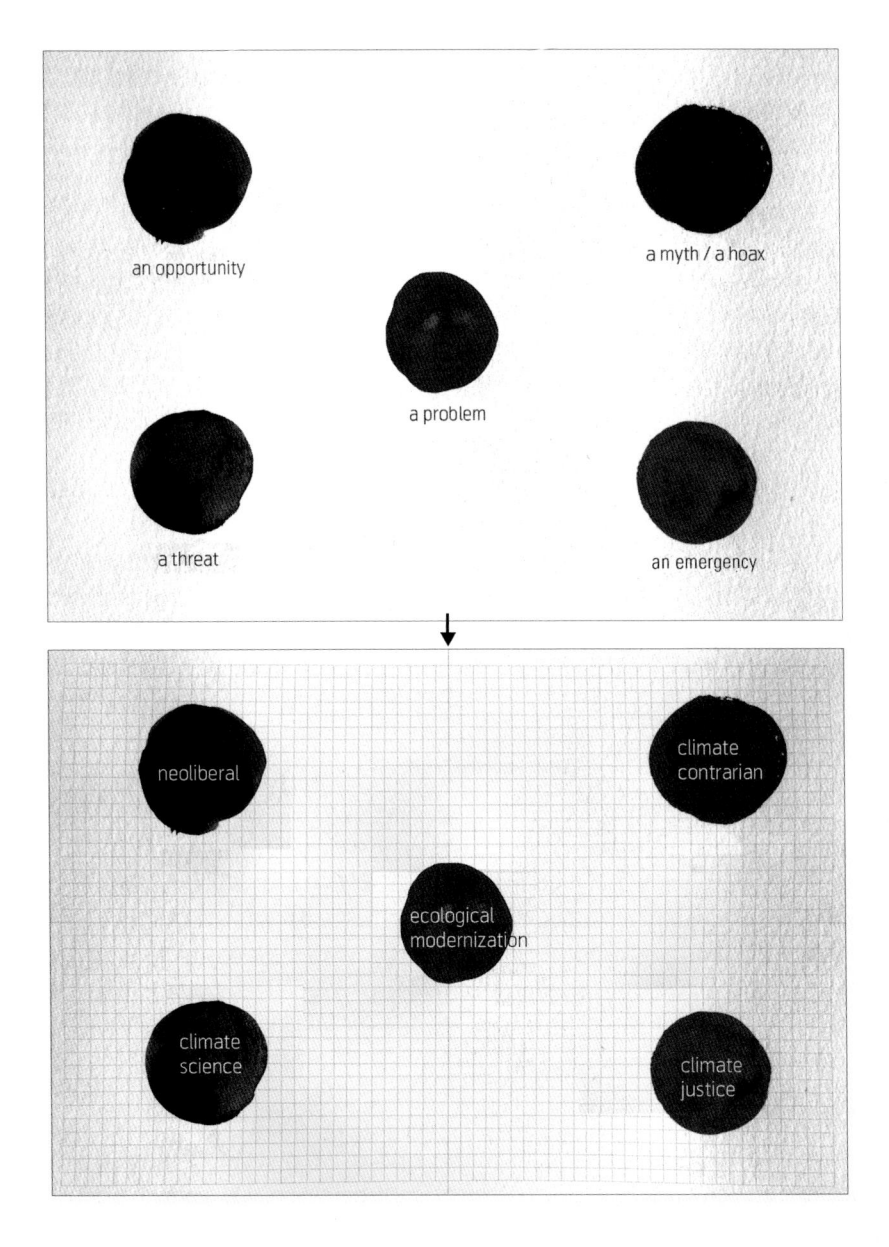

Fig. 15.6 Five Frames and Five Discourses on the Matrix – sketches for *Network of Actors*. Boehnert, 2014.

need for dramatic emissions reduction (climate science, climate justice and ecological modernization) must be aware of the ways in which the neoliberal discourse appropriates their rhetorical positions. Neoliberal actors often use the language of the environmental movement to gain and maintain legitimacy and public trust. The danger here is that the climate movement's work in creating awareness and policy opinions responding to climate change is simply used as convenient rhetoric and public relations messaging for continued and even exacerbated carbon intensive development. Discursive confusion results from a

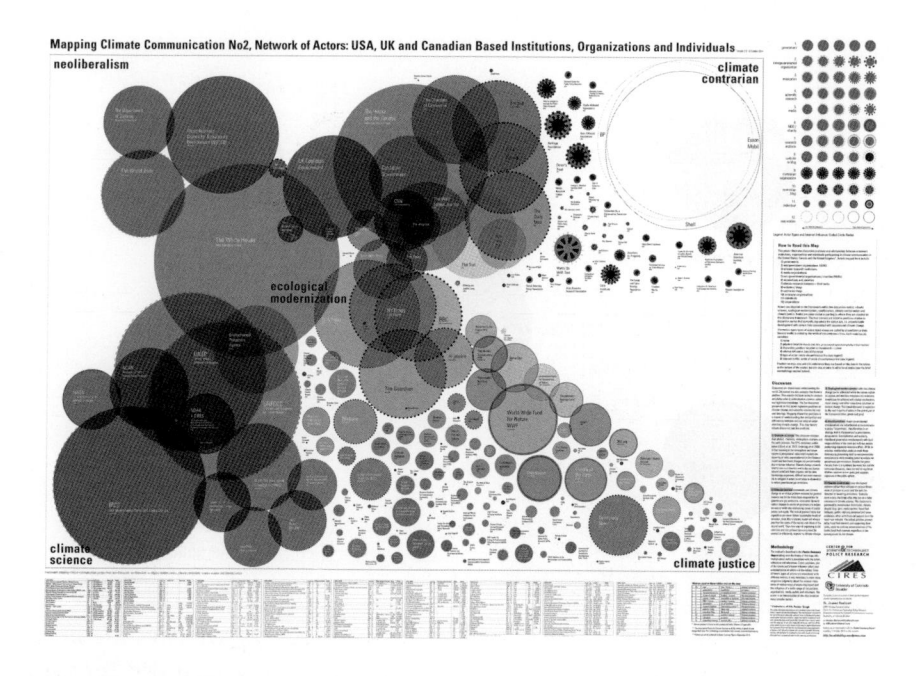

Fig. 15.7 *Mapping Climate Communication, No. 2 Network of Actors*, Boehnert, 2014.

situation where governing institutions want to appear to be providing responsible leadership in the face of the risks associated with climate change. But since effective policy responses to climate imperatives are difficult or even impossible within the ideological scaffolding of the neoliberal political project, the rhetoric is not reflect in policy decisions. Discursive confusion describes the difference between rhetoric and action on issues of climate change. Ultimately it empowers the contrarian lobby as effective climate policy is stalled. With these dynamics in mind, it is evident that contrarians are not the only ones preventing action on climate change.

Critical Information Design

With an understanding of the malleability of perception (i.e. perceptual habits are learned and culturally constructed) and the politics of representation (i.e. certain groups have more power to create representations than others) communication designers must move beyond the presentation of data alone as a foundation for analysis. Information design guru Edward Tufte claims 'excellence in presenting information requires mastering the craft and the spurning the ideology' (1990, 35). But *all communication embodies ways of knowing, perspectives and ideologies* – whether one is aware of our own ideological premises or not. The notion of one objective truth on politicized issues is suspect from an ecological perspective and also from anti-sexist, anti-racist and anti-imperialist perspectives (since historical naturalizations have been used to oppress constituencies, as in the case with early ecological theory – see p. 180). In many cases on issues of controversy, the most relevant fact is that powerful groups are able to communicate perspectives that support their own ideological commitments and material interests.

Data visualization exemplify values and worldviews that determine which data is selected, as well as the methods, media and styles that are used to communicate information. These decisions often serve political priorities. This is especially true for controversial topics – including most environmental issues. Critical approaches to information design question how communication works to reproduce or challenge power, what is being communicated, and how it is communicated. The concepts of digital positivism, datawash and darkdata are building blocks for critical information design literacy. Power and ideology can hide behind the presentation of data such that some interests are presented and others are masked. Articulating the limits of data visualization makes space for more refined approaches such as knowledge visualization where information is presented in context, thereby encouraging more integrated and actionable learning.

Notes

1. This policy is now changing.
2. In many cases 'unintended consequences' can be anticipated if institutional resources are put into investigating risks.
3. I developed the discourse mapping method as part of my *Mapping Climate Communication* research during my CIRES Visiting Research Fellowship at the Center for Science and Technology Policy Research, University of Colorado Boulder. For more information see *Mapping Climate Communication: Poster Summary Report* (Boehnert 2014b).

Conclusion: Towards the Ecocene

Design is a practice primed to make sustainability not only possible but appealing. A new economy mindful of ecological limits must become the desirable option. Design is involved with shifting and re-signifying dominant norms, meanings and aspirations. With prefigurative practices and new social imaginaries, design can help make alternative ways of living possible. Guattari evokes an ecological revolution that takes into account 'domains of sensibility, intelligence and desire' (2000, 20). To do this work we must 'confront capitalism's effects in the domain of mental ecology in everyday life' (Ibid., 33). Capitalism depends on cultural industries to nurture subjectivities. Designers regularly create artefacts, interactions and communication that encourage users and audiences to do things in new ways. With this expertise in influencing subjectivities and behaviours, design is poised to facilitate social change – once designers move beyond their comfort zones and disciplinary boundaries to meet the complexity of environmental and social challenges.

Ecological literacy provides a basis for integrated thinking about sustainability and nurtures a frame of mind that prioritizes ecological imperatives. It also politicizes design processes. Confronting environmental problems involves engaging with the political challenges necessary to transform system structures and industrial systems. Creating ecologically viable alternatives often requires intervening with the unsustainable and challenging the forces that de-prioritize ecological values. The work of building new social practices, organizations and institutions requires critical, participatory and political design. Designers can engage with and learn from social movements who have a legacy of creating agency and developing means to challenge cultural practices and institutions. This endeavour is a social learning process that includes unlearning damaging behaviour patterns, values and aspirations. Questions for those concerned with sustainability include: should profit-making for those with capital be the most important priority in the social and political order? If not what can we do about this? The fact that these conversations are still marginal despite serious environmental threats is an issue that must also be interrogated. The scale of change necessary to address climate change alone demands dramatic transformations. Powerbrokers typically favour weak sustainability discourses and it will take concerted efforts to dislodge these narratives.

Within the context of an increasingly visual culture, visual communication can facilitate ecological learning. Due to the fact that representations of the environment are highly contested, the visual communication of ecological information must sit within an informed critique of relevant power dynamics – including the production and reproduction of various environmental discourses. Revealing the impacts of industrial systems and consumer culture on the environment often involves confrontations with powerful vested interests and

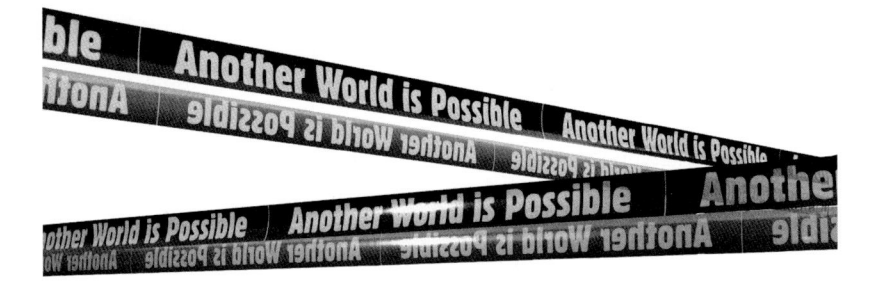

Fig. C.1 *Another World is Possible.*

cultures of denial. Like all matters in regards to sustainability, the onus of using intellectual and other resources towards the well-being of people and planet sits at odds with short-term profit incentives.

Sustainability learning must be part of formal design education. Educational establishments have a responsibility to ensure that students graduate with an understanding of the consequences of unsustainable ways of living, the knowledge they need to make ecologically informed decisions and the skills and capacities to do something about it. Bridging the value/action gap remains a challenge for sustainability communicators, designers and educators alike. There remains a great distance between accepting something as an intellectual truth and perceiving, thinking and acting according to this position. Apathy can be bred by institutional practices that fail to practice what they preach: 'students learn that it is sufficient only to learn about injustice and ecological deterioration without having to do much about them, which is to say, the lesson of hypocrisy' (Orr 1992, 104). Design schools need to walk their talk in regard to sustainability in the curriculum and in institutional practice and policy. Another world is possible once enough people are prepared to do the work to make it happen.

Crisis as Catalyst

Crises are unstable periods that can also be turning points. The instability generated in these periods can bring about dramatic change. Right-wing economic theorist Milton Friedman claims:

> Only a crisis – actual or perceived – produces real change. When that crisis occurs, the actions that are taken depend on the ideas that are lying around. That, I believe, is our basic function: to develop alternatives to existing policies, to keep them alive and available until the politically impossible becomes the politically inevitable. (1982, ix)

On the other side of the political spectrum, social justice writer Naomi Klein describes how power regimes use (and sometimes even manufacture) crisis conditions to enforce unjust policies on shocked populations to support the interests of elites – including dramatic appropriations of public wealth (2008, 2014). Crises are also times when social movements can make the politically impossible inevitable. The recent economic crisis was hailed by some as the end of neoliberal hegemony. It presented an opportunity to intervene but instead the post-2009 period saw further concentration in wealth in the top percentile: the richest 1,000 families in the UK have seen their wealth more than double since the 2009 recession (Garside 2015). The next economic crisis will provide similar opportunities. The process of developing ecological thought to a sufficient state that it can be used to redesign system structures and ways of living once the collective illusion of radical discontinuity from nature is shattered is a substantial task. Design can respond by working on long-term, ongoing strategies of intervention and renewal. With a focus on regenerative, qualitative and sustainable distributed prosperity designers address crises that ecoist mindsets have generated.

The stresses in the three domains (ecological, social and psychological) reflect a paradigm in tatters, no longer capable of explaining or addressing current problems. Ecoism is embedded into system structures and worldviews. This paradigm is held onto with brutish determination by the ruling order. The current order's values and priorities are not fit for purpose. The Cartesian worldview, a relic of ecoist, sexist, racist and colonialist assumptions, no longer works to interpret and respond to contemporary problems. Escalating eco-social crises will continue until we adapt. Guattari maintains:

> The Earth is undergoing a period of intense techno-scientific transformations. If no remedy is found, the ecological disequilibrium this has generated will ultimately threaten the continuation of life on the planet's surface. Alongside these upheavals, human modes of live, both individual and collective, are progressively deteriorating. (2000, 19)

For many people, this is a harsh process. There is no easy solution. The emergence of a potential future Ecocene depends on humankind's willingness and capacity to organize, congenial with ecological context, to higher levels of complexity with respect for our ecological context and each other. Coordinated and cooperative action is where all power, ingenuity and genuine social progress lie. This will be achieved, in part, by reflectively designing appropriate responses informed by ecological literacy.

It is also possible that the Earth will descend into decreasing levels of ecological complexity. This shrinking is already apparent in the rapid extinction of higher mammals, growing deserts, shrinking rainforests and disappearing cultures and peoples. This trajectory will only be averted with multidimensional work by on ecological ways of knowing – including ecological epistemology, rationality, ethics, perception and identity. Ecological theory offers a basis for renewal. For ecological theorists it is clear that 'the closer the economy approaches the scale of the whole Earth the more it will have to conform to the physical behaviour mode of the Earth' (Daly 2008, 1). Once enough people recognize this fact and act strategically on this knowledge, the Ecocene will become possible.

Transitions will involve both small steps and dramatic phase shifts. Design writer John Thackara refers to the 'Great Turning' (see p. 142) as a fundamental shift in our understanding of our relationship with the planet. He describes how this vision of 'reawakening' informs a current

> quietly unfolding transformation ... consistent with the way scientists, too, explain systems change. By their account, a variety of changes, interventions, and disruptions accumulate across time until the system reaches a tipping point: then, at a moment cannot be predicted, a small release of energy triggers a much larger release, or phase shift, and the system as a whole transforms. Sustainability, in other words, is not something that to be engineered, or demanded by politicians; it's a condition that emerges through incremental as well as abrupt change at many different scales. (2015a, 9)

People around the globe are working towards this shift to viable, sustainable futures. We cannot know if we will be successful but we do know we will fail if we don't try – if we continue along the dominant current trajectory. Meanwhile the environment itself will survive (in diminished form) whatever humankind does to the ecological conditions that sustain civilization. But there is no alternative for humanity but to design in sync with the ecological context that sustains us.

Climate change and other environmental problems demand shifts on an order of magnitude well beyond the trajectory of business-as-usual. Ecomodernists' fantasies on technological salvation are unhelpful distractions when they sideline the ecological in sensibilities, in theory, in transformative practice and in the multiple material struggles faced by those fighting environmental injustices. The primary identification must be the ecological. We start and end our lives as part of our ecological context. We are principally biological, ecological beings with sentience and even soul. Meanwhile, the design industry is focused on the next season's fonts. These same skills could be used for different ends. Designers' ability to make the shape of a letter so meaningful, appealing and even part of our identity can be redirected to do the significant work of making viable ways of living over time possible. We can harness this capacity to embody the spirit of an idea in material form to make something consequential.

References

Accounting for Sustainability (2012) *Valuing Natural Capital: The Economic Invisibility of Nature Images*. Accessed online: http://www.accountingforsustainability.org/ embedding-sustainability/the-economic-invisibility-of-nature-information-is-beautiful-images
Albrecht G (2012) Psychoterratic Conditions in a Scientific and Technological World. In P H Kahn, Jr & P H Hasbach (eds.) *Ecopsychology: Science, Totems and the Technological Species*. London: MIT Press, 241–264.
Albrecht G (2013) Psychoterratica: Creating a Language for Our Psychoterratic Emotions and Feelings. Accessed online: psyche {mind}; terra {earth} http://www.psychoterratica.com/more.html
Ansari A, Abdulla D, Canli E, Keshavarz M, Kiem M, Oliveira P, Prado L, & Schultz T (2016) Decolonising Design: Editorial Statement. 27 June 2016. Accessed online: http://www.decolonisingdesign.com/general/2016/editorial/
Armstrong R (2014) 3D Printing Will Destroy the World Unless It Tackles the Issue of Materiality. *The Architectural Review*, 31 January.
Armstrong R (2015) [Keynote presentation] *Urban Ecologies*. Toronto, Canada: OCAD, 18 June.
Arsel M & Buscher B (2012) Nature™ Inc.: Changes and Continuities in Neoliberal Conservation and Market-Based Environmental Policy. *Development and Change*, 43 (1), 53–78.
Assadourian E (2010) The Rise and Fall of Consumer Culture. In World Watch Institute (ed.) *2010 State of the World: Transforming Cultures*. London: W W Norton & Company.
Austin A, Barnard J, & Hutcheon N (2014), *Advertising Expenditure Forecasts: December 2014*. London: Zenith.
Austin A, Barnard J, & Hutcheon N (2016), *Advertising Expenditure Forecasts: December 2016*. London: Zenith.
Ban K M (2009) Shaping the Climate Change Message, [speech] World Economic Forum Davos: 29 January 2009. Accessed online: https://goo.gl/VLUYL9
Barabasi A L (2003) *Linked*. London: Plume
Barry A M (1997) *Visual Intelligence: Perception, Image, and Manipulation in Visual Communication*. Albany, NY: SUNY Press.
Bateson G (1972) *Steps to an Ecology of Mind*. Chicago, IL: University of Chicago Press.
Bateson G (1980) *Mind and Nature*. London: Bantam Books.
Bateson G & Bateson M C (1988) *Angels Fear: Towards an Epistemology of the Sacred*. Chicago, IL: University of Chicago Press.
Bateson M C (2005) *Our Own Metaphor*. Cresskill, NY: Hampton Press, The Institute for Cultural Studies.
Battiste M (2002) Indigenous Knowledge and Pedagogy in First Nations Education: A Literature Review with Recommendations. Prepared for the National Working Group on Education and the Minister of Indian Affairs and Northern Affairs Canada (INAC).
Bennett J (2010) *Vibrant Matter: A Political Ecology of Things*. Durham, NC: Duke University Press.

Benson E & Perullo Y (2017) *Design to Renourish: Sustainable Graphic Design in Practice*. London: CRC Press, Taylor and Francis Group.
Benyus J M (1997) *Biomimicry*. New York, NY: HarperCollins.
Berger J (2008/1972) *Ways of Seeing*. London: BBC and Penguin Books.
Bertalanfry L (1968) *General System Theory*. New York, NY: George Braziller.
Biehl J (1991) *Finding Our Way: Rethinking Ecofeminist Politics*. Montreal: Black Rose Books.
Binfield K (2004) *Writings of the Luddites*. Baltimore, MD: Johns Hopkins University Press.
Bliss S (2016) Ecomodernism & Degrowth. 5th Degrowth Conference, Budapest, 30 August–3 September.
Bliss S & Kallis G (2017) *Kiss Nature Goodbye. The Dangerous Illusions of Post-environmentalism*. Manuscript in preparation.
Block F (2001) Introduction in K Polanyi [1944] *The Great Transformation*. Boston, MA: Beacon Press.
Boehnert J (2012) *The Visual Communication of Ecological Literacy: Designing, Learning and Emergent Ecological Perception*. PhD thesis, Department of Design, University of Brighton.
Boehnert J (2014a) Ecological Perception: Seeing Systems. In Proceedings of *DRS 2014: Design's Big Debates*, Umea, Sweden, 16–19 June.
Boehnert J (2014b) *Mapping Climate Communication: Poster Summary Report*. Center for Science and Technology Policy Research, CIRES, University of Colorado Boulder.
Boehnert J (2015) The Politics of Data Visualisation. *Discover Society*, 3 August 2015, 23.
Boehnert J (2016) Data Visualisation Does Political Things. In Proceedings of *DRS 2014: Design + Research + Society: Future Focused Thinking*, Brighton, UK, 27–30 June.
Boehnert J (2017) Ecological Theory in Design: Participant Designers in an Age of Entanglement. In R B Egenhoefer (ed.) *Routledge Handbook of Sustainable Design*. New York, NY: Routledge.
Bohm D (1992) *Thought as a System*. London: Routledge.
Bollier D (2016) Seeing Wetiko. [blog] David Bollier: News and Perspectives on the Commons. Accessed online: http://bollier.org/blog/seeing-wetiko
Bond K (2017) Energy Return on Investment: The Dawn of the Age of Solar. *Trusted Sources*. 8 March. Accessed online: http://www.trustedsources.co.uk/blogs/new-energy/energy-return-on-investment-the-dawn-of-the-age-of-solar
Bookchin M (1993) *Defending the Earth: Debate Between Murray Bookchin and Dave Foreman*. New York, NY: Black Rose Books, 129–130.
Bourdieu P (2001) *Masculine Domination*. Translated by R Nice. London: Polity.
Bourdieu P (2010/1987) *Distinction: A Social Critique of the Judgement of Taste*. Translated by R Nice. London: Routledge.
Bower C (2005/1999) The Role of Education and Ideology in the Transition from a Modern to a More Bioregionally Oriented Culture. In M V McGinnis (ed.) *Bioregionalism*. London: Routledge.
Brand U & Vadrot A (2013) Epistemic Selectivities and the Valorisation of Nature: The Cases of the Nagoya Protocol and the Intergovernmental Science-Policy Platform for Biodiversity and Ecosystem Services (IPBES). *Law, Environment and Development Journal*, 9, 202–222.
Braungart M & McDonough W (2002) *Cradle to Cradle*. New York, NY: North Point Press.
Brooks D (2013) What Data Can't Do. *The New York Times*. 19 February.
Brown L (2008) *Plan B 3.0 – Mobilizing to Save Civilization*. New York, NY: W W Norton & Co.
Brulle R B (2014) Institutionalizing Delay: Foundation Funding and the Creation of U.S. Climate Change Counter-Movement Organizations. *Climatic Change*, 122 (4), 681–694.
Brundtland G H et al. (1987) *Our Common Future. Report of the World Commission on Environment and Development* (WCED) [known as the *Brundtland Report*]. Oxford: Oxford University Press.

Buchanan R (1992) Wicked Problems in Design Thinking. *Design Issues*, Vol. 8, No. 2 (Spring, 1992), MIT Press, 5–21.
Bullard R D (ed.) (1990) *Dumping in Dixie: Race, Class, and Environmental Quality*. Boulder, CO: Westview Press.
Burns C, Cottam H, Vanstone C, & Winhall J (2006) *RED Paper 02: Transformation Design*. London: Design Council.
Capra F (1982) *The Turning Point: Science, Society and the Rising Culture*. New York, NY: Simon & Schuster.
Capra F (1997) *The Web of Life*. London: HarperCollins.
Capra F (2003) *The Hidden Connections*. London: Flamingo.
Capra F (2005) Speaking Nature's Language: Principles of Sustainability. In Z Barlow & M K Stone (eds.) *Ecological literacy*. San Francisco, CA: Sierra Club Books.
Capra F & Henderson H (2009) *Qualitative Growth*. London: The Institute of Chartered Accountants in England and Wales.
Caradonna et al. (2015) A Degrowth Response to the Ecomodernist Manifesto. *Resilience*. 5 May. Accessed online: http://www.resilience.org/wp-content/uploads/articles/General/2015/05_May/A-Degrowth-Response-to-An-Ecomodernist-Manifesto.pdf
Carbon Brief Staff (2013) Energy Return on Investment – Which Fuels Win? *Carbon Brief*. 20 March. Accessed online: https://www.carbonbrief.org/energy-return-on-investment-which-fuels-win
Caritas Data (2016) *Top 3000 Charities 2016*, 24th edition. London: Carity Financials.
Castree N (2008) Neoliberalising Nature: The Logics of Deregulation and Reregulation. *Environment and Planning A*, 40, 131–152.
Castree N (2014) The Anthropocene and the Environmental Humanities: Extending the Conversation. *Environmental Humanities*, 5 (1), 233–260.
Centre for Alternative Technologies (2010) *Zero Carbon Britain 2030: A New Energy Strategy*. Llwyngwern, Machynlleth, Powys, UK: Centre for Alternative Technology.
Centre for Alternative Technologies (2015) *Zero Carbon: Making It Happen*. Llwyngwern, Machynlleth, Powys, UK: Centre for Alternative Technology.
Chabris C F & Kosslyn S M (2005) Representational Correspondence as a Basic Principle of Diagram Design. In S O Tergan & T Keller (eds.) *Knowledge and Information Visualization*, LNCS 3426, 36–57.
Chapman J (2005) *Emotionally Durable Design*. London: Earthscan.
Chomsky N (2013a) *On Anarchism*. London: Penguin Books.
Chomsky N (2013b) Noam Chomsky Slams Canada's Shale Gas Energy Plan. *The Guardian*. Friday 1 November. Accessed online: https://www.theguardian.com/ environment/2013/nov/01/noam-chomsky-canadas-shale-gas-energy-tar-sands
Cohen P (2016) A Bigger Economic Pie, but a Smaller Slice for Half of the U.S. *The New York Times*. 6 December. Accessed online: http://tiny.cc/709jiy
Cohen S (2001) *States of Denial: Knowing about Atrocities and Suffering*. Cambridge, MA: Polity Press.
Common Cause (2016) [website]. Common Cause. Machynlleth, Powys, Wales: Common Cause Foundation. Accessed online: http://valuesandframes.org/handbook/2-how-values-work
Commoner B (1971) *The Closing Circle*. London: Jonathan Cape.
Connolly W (2013) *The Fragility of Things*. London: Duke University Press.
Corby T (2015) From Un-data to 'Un-visualisation, Transforming Data: Creative and Critical Directions in the Arts and Humanities. University of Westminster, 24 October 2015.
Cox R (May 2007) Nature's "Crisis Disciplines": Does Environmental Communication Have an Ethical Duty? *Environmental Communication: A Journal of Nature and Culture*, 1 (1), 5–20.

Credit Suisse: Shorrocks A, Davies J, Lluberas R & Koutsoukis A (2016) *Global Wealth Report 2016*. Zurich: Credit Swisse AG. Accessed online: http://tiny.cc/6qbkiy

Crompton T (2008) *Weathercocks and Signposts*. Godalming, Surrey: WWF-UK.

Crompton T (2010) *Common Cause: The Case for Working with Our Cultural Values*. Godalming, Surrey: WWF-UK.

Crompton T (June 2013) *Putting a Price on the Priceless: Valuing Nature?* Presentation at Royal Geographical Society, London. [Event] Accessed online: https://soundcloud.com/britishecologicalsociety/valuing-nature-18-june–2013

Crompton T & Kasser T (2009) *Meeting Environmental Challenges: The Role of Human Identity*. Godalming: WWF.

Cross N (1990) The Nature and Nurture of Design Ability. *Design Studies*, 11 (3), 127–140.

Cross N (2010) Design Thinking as a Form of Intelligence. Proceeding from Design Thinking and Research Symposium Eight 2010. DTRS8 2010. Sydney, 19–20 October 2010.

Crutzen P J & Stoermer E F (2000) *Global Change Newsletter*. International Geosphere–Biosphere Programme (IGBP), 41.

Cullen J M & Allwood J M (2010) The Efficient Use of Energy: Tracing the Global Glow of Energy from Fuel to Service. *Energy Policy*, 38 (1), 75–81.

Curtis A (2011) The Use and Abuse of Vegetational Concepts. *All Watched over by Machines of Loving Grace*, Part Two. [Documentaries series: television programme] BBC Two. 23 May 2011.

D'Alisa G, Demaria F, & Kallis G (eds.) (2014) *Degrowth: A Vocabulary for a New Era*. London: Routledge.

Daly H (1972) *The Steady State Economy*. London: W H Freeman and Co Ltd.

Daly H (1996) *Beyond Growth*. Washington, DC: Beacon Press.

Daly H (2008a) *A Steady-State Economy*. London: Sustainable Development Commission.

Daly H (2008b) Big Idea: A Steady State Economy. *Adbusters*. 17 December 2008. Accessed online: http://www.adbusters.org/article/big-idea-a-steady-state-economy/

Dardot P & Laval C (2013) *The New Way of the World: On Neoliberal Society*. London: Verso.

Deregowski J B (1980) *Illusions, Patterns and Pictures: A Cross-Cultural Perspective*. London: Academic Press.

D'Ignazio C (2015) What Would Feminist Data Visualisation Look Like? *MIT Center for Civic Media*. Accessed online: https://civic.mit.edu/feminist-data-visualization

Dondis D A (1973) *A Primer of Visual Literacy*. Cambridge, MA: MIT Press.

Doyle J (2009) Seeing the Climate? The Problematic Status of Visual Evidence in Climate Change Campaigning. In S Dobrin & S Morey (eds.) *Ecosee: Image, Rhetoric, and Nature*. New York: State University of New York Press, 279–298.

Dryzek J S (2005) *The Politics of the Earth*, 2nd edition. Oxford: Oxford University Press.

Dryzek J S (2013) *The Politics of the Earth*, 3rd edition. Oxford: Oxford University Press.

Dunne A & Raby F (2013) *Speculative Everything: Design, Fiction, and Social Dreaming*. Cambridge, MA: MIT Press.

Edwards R & Usher R (1994) *Postmodernism and Education*. London: Routledge

Ehrenfeld J (2008) *Sustainability by Design*. London: Yale University Press.

Eisler R (1988) *The Chalice & the Blade*. New York, NY: HarperCollins Publishers.

EMAPS, Electronic Maps to Assist Public Science (2014) *Climaps by Emaps: A Global Issue Atlas of Climate Change Adaptation*. Accessed online: http://climaps.eu/#!/home

Erle et al. (2012) Used Planet: A Global History. *PNAS*, 110 (20), 7978–7985.

Ewen S (2003/1990) Notes for the New Millennium. Is the Role of Design to Glorify Corporate Power? Is the Role of Design to Glorify Corporate Power? In S Heller & V Vienne (eds.) *Citizen Designer*. New York, NY: Allworth Press. 191–195.

FAO (2014a) *Family Farmers: Feeding the World, Caring for the Earth*. Rome: Food and Agriculture Organization of the United Nations.
FAO (2014b) *The State of Food and Agriculture 2014 in Brief*. Rome: Food and Agriculture Organization of the United Nations.
Fuad-Luke A (2009) *Design Activism*. London: Earthscan.
Fauset C (2008) *Technofixes: A Critical Guide to Climate Change Technologies*. London: Corporate Watch.
Feibleman J K (1954) Theory of Integrative Levels. *The British Journal for the Philosophy of Science*, 5, 59–66.
Fioramonti L (2013) *Gross Domestic Problem*. London: Zed Books.
Fisher A (2012) What is Ecopsychology? A Radical View. In P H Kahn & P Hasbach (eds.) *Ecopsychology: Science, Totems, and the Technological Species*. London: The MIT Press, 79–114.
Flavin C (2010) Preface. In L Starke & L Mastny (eds.) *State of the World: Transforming Cultures from Consumerism to Sustainability*. New York, NY: W W Norton & Company.
Forceville C (2008) Metaphor in Pictures and Multimodal Representations. In R Gibbs (ed.) *The Cambridge Handbook of Metaphor and Thought*. Cambridge: Cambridge University Press.
Foster J B (2002) *Ecology Against Capitalism*. New York, NY: Monthly Review Press.
Foucault M (1972/2002) *The Archaeology of Knowledge & The Discourse of Knowledge*. Translated by A Smith. London: Routledge Classics.
Foucault M (1980) *Power/Knowledge: Selected Interviews and Other Writings 1972–1977*. Edited by C Gordon. New York, NY: Pantheon Books.
Frankle V (1959) *Man's Search for Meaning*. Boston, MA: Beacon Press.
Frazier C M (2016) Troubling Ecology: Wangechi Mutu, Octavia Butler, and Black Feminist Interventions in Environmentalism. *Critical Ethnic Studies*, 2 (1), 40–77.
Friedman M (1982) *Capitalism and Freedom*. Chicago, IL: University of Chicago Press.
Fromme E (1964) *The Heart of Man, Its Genius for Good and Evil*. Herndon, VA: Lantern Books.
Fry T (2009) *Design Futuring*. Oxford: Berg.
Funtowicz S & Ravetz J (2003) Post-normal Science. In E Neumyer (ed.) *Internet Encyclopaedia of Ecological Economics*, International Society for Ecological Economics.
Gardner H (1983) *Frames of Mind: The Theory of Multiple Intelligences*. New York, NY: Basic Books.
Garland K et al. (1964) *The First Things First Manifesto*, 1st edition. London: Goodwin Press.
Garland K et al. (1999) The First Things First Manifesto 2000. *Eye no. 33 (9)* London: *Eye Magazine*, and New York: *AIGA Journal of Graphic Design*.
Garside J (2015) Recession Rich: Britain's Wealthiest Double Net Worth since Crisis. *The Guardian*. 26 April.
Gee T (2011) *Counter Power*. Oxford: New Internationalist Publications.
Georgescu-Roegen N (1972) *The Entropy Law and the Economic Process*. Cambridge, MA: Harvard University Press.
Giddens A (1990) *The Consequences of Modernity*. Stanford, CA: Stanford University Press.
Gilbert J (2014) *Common Ground: Democracy and Collectivity in an Age of Individualism*. London: Pluto Press.
Gilpin L (2014) The Dark Side of 3D Printing: 10 Things to Watch. *TechRepublic*. 5 March. Accessed online: http://www.techrepublic.com/article/the-dark-side-of-3d-printing-10-things-to-watch/
Global Footprint Network (2016) Footprint Basics – Overview. Accessed online: http://www.footprintnetwork.org /en/index.php/gfn/page/footprint_basics_overview)

Global Footprint Network (2017) Glossary. Accessed online: http://www.footprintnetwork.org/resources/glossary/
Global Witness (2014) *Deadly Environment: The Dramatic Rise in Killings of Environmental and Land Defenders*. London: Global Witness.
Global Witness (2015) *How Many More? 2014's Deadly Environment: The Killing and Intimidation of Environmental and Land Activists, with a Spotlight on Honduras*. London: Global Witness.
Goleman D (1995) *Emotional Intelligence*. New York, NY: Bantam Books.
Goleman D (2009) *Ecological Intelligence: How Knowing the Hidden Impacts of What We Buy Can Change Everything*. New York, NY: Broadway Books.
Goodbun J C (2011) *The Architecture of the Extended Mind*. Doctoral degree at the University of Westminster. June 2011.
Graeber D (2000) Give It Away. *In These Times.com*, 24 (19), August 21.
Graeber D (2002) The New Anarchists. *New Left Review 13*. January–February, 61–73.
Graeber D (2011) *Debt: The First 5000 Years*. Brooklyn, NY: Melville House.
Gray L (1995) Shamanic Counseling and Ecopsycholgy. In T Roszak, M Gomes, & A Kanner (eds.) *Ecopsychology*. San Francisco, CA: Sierra Club Books, 172–182.
Griffin D R (1992) Introduction to Suny Series in Construction Postmodern Thought. In D Orr (ed.) *Ecological Literacy*. Albany, NY: Suny Press.
Guattari F (2000/1989) *The Three Ecologies*. Translated by I Pindar & P Simon. London: Continuum.
Haeckel E (1866) *General Morphology of Organisms; General Outlines of the Science of Organic Forms Based on Mechanical Principles Through the Theory of Descent as Reformed by Charles Darwin*. Berlin.
Hall A S, Lambert J G, & Balogh S B (2014) EROI of Different Fuels and the Implications for Society. *Energy Policy*, 64, January, 141–152.
Hall S (1973) *Encoding and Decoding in the Television Discourse*. Birmingham, UK: Centre for Cultural Studies, University of Birmingham.
Hamilton C (2015) An App for an Ailing Planet. *New Philosopher*. Issue #6: progress. 16 February 2015. Accessed online: https://www.newphilosopher.com/articles/an-app-for-an-ailing-planet/
Hansen J (2011) It's a Hard-Knock Butterfly's Life. Can a Lady Monarch Provide a Role Model? Accessed online: www.columbia.edu/~jeh1/mailings/ 2011/20110928_Butterfly.pdf
Hansen J et al. (2007) Climate Change and Trace Gases. *Philosophical Transactions of the Royal Society*, A, 365, 1925–1954.
Haraway D (1988) Situated Knowledges: The Science Question in Feminism and the Privilege of Partial Perspective. *Feminist Studies*, 14 (3), 575–599.
Haraway D (2014) [Keynote presentation] Anthropocene, Capitalocene, Chthulucene: Staying with the Trouble. 5 September 2014. Accessed online: https://vimeo.com/97663518
Haraway D (2015) Anthropocene, Capitalocene, Plantationocene, Chthulucene: Making Kin. *Environmental Humanities*, 6, 159–165.
Harding S G (1986) *The Science Question in Feminism*. Ithaca, NY: Cornell University Press.
Harding S G (2006) *Animate Earth*. Dartington, UK: Green Books
Harper S (1995) The Way of Wilderness. In T Roszak, M Gomes, & A Kanner (eds.) *Ecopsychology*. San Francisco, CA: Sierra Club Books, 183–200.
Harris L D & Wasilewski J (2004a) Indigeneity, an Alternative Worldview: Four R's (Relationship, Responsibility, Reciprocity, Redistribution) vs 2 P's (Power & Profit). *Systems Research and Behavioral Science*, 21, 489–503.
Harris L D & Wasilewski J (2004b) Indigenous Wisdom of the People Forum: Strategies for Expanding a Web of Transnational Indigenous Interactions. *Systems Research and Behavioral Science*, 21, 505–514.

Hedges C (2009) *The Empire of Illusion*. New York, NY: Nation Books.
Hendry J (2014) *Science and Sustainability: Learning from Indigenous Wisdom*. London: Palgrave MacMillan.
Heron J (1992) *Feeling and Personhood. Psychology in Another Key*. London: Sage.
Holmes T, Blackmore E, Hawkins R, & Wakeford T (2011) *The Common Cause Handbook*. Machynlleth, Wales: Public Interest Research Centre (PIRC).
Hoornweg D, Bhada-Tata P, & Kennedy C (2013) Environment: Waste Production Must Peak This Century. *Nature*. 30 October 2013. Accessed online: http://www.nature.com/news/environment-waste-production-must-peak-this-century-1.14032
Horn R (1998) *Visual Language: Global Communication for the 21st Century*. Brainbridge Island, Washington, DC: Macro VU Press.
Horn R (2001) Knowledge Mapping for Complex Social Messes. A Presentation at 'Foundations in the Knowledge Economy, 16 July 2001. Retrieved 8 November 2013, from http://www.stanford.edu/~rhorn/SpchPackard.html
Horn R (2005) What We Do Not Know: Using Information Murals to Portray Scientific Ignorance. *Futures*, 38 (3), 372–377.
Hughes L (2011) *The No-Nonsense Guide to Indigenous People*. Cambridge: New Internationalist.
Hughes L (2012) *The No-Nonsense Guide to Indigenous Peoples*. Oxford: New Internationalist.
Huesemann M & Huesemann J (2011) *Techno-Fix: Why Technology Won't Save Us or the Environment*. Gabriola Island, BC, CAN: New Society Publishers.
Humantific (2010) SenseMaking for ChangeMaking. Icograda's Straight to Business Conference. Madrid, June 2010. New York: Humanific Publications. Accessed online: http://issuu.com/humantific/docs/humantific_2010
Inman M (2013) How to Measure the True Cost of Fossil Fuels. *Scientific American*. 1 April. Accessed online: http://www.scientificamerican.com/article/how-to-measure-true-cost-fossil-fuels/
Irwin T (2015) Transition Design: A Proposal for a New Area of Design Practice, Study, and Research. *Design and Culture*, 7 (2), 229–246.
Irwin T, Kossoff G, Tonkinwise C, & Scupelli P (2015) *Transition Design 2015*. Pittsburgh, PA: Carnegie Mellon University.
Irwin T, Tonkinwise C, & Kossoff G (2013) Transition Design: Re-conceptualizing Whole Lifestyles. *Head, Heart, Hand: AIGA Design Conference*, Minneapolis, 12 October 2013. Accessed online: http://www.aiga.org/video-HHH-2013-irwin-kossoff-tonkinwise
Ison R (2008) Systems Thinking and Practice for Action Research. In P Reason & H Bradbury (eds.) *The SAGE Handbook of Action Research*. London: Sage.
Jackson M C (1990) Critical Systems Thinking: Beyond the Fragments. *System Dynamics Review*, 10 (2–3), 213–229.
Jackson T (2009) *Prosperity Without Growth?* London: Sustainable Development Commission.
Jacques P (2005) Ecology as Resistance. *Peace Review: A Journal of Social Justice*, 17, 435–441.
Kahan D M (2010) Fixing the Communications Failure. *Nature*, 463, 292–297.
Kahn R (2010) *Critical Pedagogy, Ecological Literacy, and Planetary Crisis*. New York, NY: Peter Lang.
Kallis G (2015) An Ecomodernist Mishmash. Eco-modernization Is an Oxymoron. *Degrowth*. Accessed online: https://www.degrowth.de/en/2015/05/an-ecomodernist-mishmash/
Kanner A D & Gomes M E (1995) The All-Consuming Self. In T Roszak, M Gomes, & A Kanner (eds.) *Ecopsychology*. San Francisco, CA: Sierra Club Books, 111–121.
Kari-Oca 2 Declaration (17 June 2012) Indigenous Peoples global conference on Rio+20 and Mother Earth. Accepted by acclamation, Kari-Oka Village, at Sacred Kari-Oka Púku, Rio de Janeiro, Brazil. Accessed online: http://indigenous4mother earthrioplus20.org/ kari-oca-2-declaration/

Kimmerer R W (2015) Nature Needs a New Pronoun: To Stop the Age of Extinction, Let's Start by Ditching "It". *Yes Magazine*. 30 March 2015. Accessed online: http://www.yesmagazine.org/issues/together-with-earth/alternative-grammar-a-new-language-of-kinship

King D (2015a) Exposing Technocracy – The Mindset of Industrial Capitalism. *The Ecologist*. 27 June. Accessed online: http://tiny.cc/p4kkiy

King D (2015b) Our Heritage, the Luddite Rebellion 1811–1813. The Luddites at 200 for Action Against Technology 'Hurtful to Commonality'. Accessed online: http://www.luddites200.org.uk/theLuddites.html

Klein N (2008) *The Shock Doctrine*. Toronto: Random House.

Klein N (2014) *This Changes Everything*. Toronto: Simon and Schuster.

Kolbert E (2014) *The Sixth Extinction: An Unnatural History*. New York, NY: Bloomsbury.

Korten D (2006a) *The Great Turning: From Empire to Earth Community*. Boulder, CO: Lynne Rienner Publishers.

Korten D (2006b) The Great Turning: From Empire to Earth Community. *YES! A Journal of Positive Futures*. Summer. Accessed online: http://www.yesmagazine.org/pdf/38Korten.pdf

Korzybski A (1994/1933) *Science and Sanity: An Introduction to Non-Aristotelian Systems and General Semantics*, 5th edition. Englewood, NJ: Institute of General Semantics.

Kossoff G (2011) *Holism and the Reconstitution of Everyday Life: A Framework for Transition to a Sustainable Society*. PhD thesis, University of Dundee.

Kreutzer D (1995) FASTbreakTM: A Facilitation Approach to Systems Breakthroughs. In S Chawla & J Renesch (eds.) *Learning Organizations: Developing Cultures for Tomorrow's Workplace*. New York, NY: Productivity Press.

Kuhn T (1970 [1962]) *The Structure of Scientific Revolutions*. Chicago, IL: University of Chicago Press.

Kurman M & Lipson H (2013) Is Eco-Friendly 3D Printing a Myth? (Op-Ed) *Live Science*. 20 July. Accessed online: http://www.livescience.com/38323-is-3d-printing-eco-friendly.html

LaDuke W (2014) Indigenous Women Telling a New Story about Energy and Climate' presented by Inside the Greenhouse. University of Colorado, Boulder, 9 December 2014.

Lakoff G (2004) *Don't Think of an Elephant: Know Your Values and Frame the Debate*. White River Junction, VT: Chelsea Green Publishing Co.

Larson B (2011) *Metaphors for Environmental Sustainability: Redefining Our Relationship with Nature*. London: Yale University Press.

Latour B (2002) *War of the Worlds: What about Peace?* Chicago, IL: Prickly Paradigm Press.

Latour B (2014) *Anthropology at the Time of the Anthropocene - A personal View of What Is to be Studied*. 113th AAA meeting in Washington on the 5th of November 2014. Accessed online: http://www.bruno-latour.fr/sites/default/files/139-AAA-Washington.pdf

Latour B (2015) *Environmental Humanities Breakthrough Dialog, Sausalito, June 2015*. http://environmentalhumanities.dukejournals.org/content/7/1/219.full

Lauderdale P (2007) Indigenous Peoples and Environmentalism. In G Anderson & K Herr (eds.) *The Encyclopedia of Activism and Social Justice*. London: Sage Publications.

Lauderdale P (2008) Indigenous Peoples in the Face of Globalization. *American Behavioral Scientist*, 51 (12), 1836–1843.

Lenzen M (2008) Life Cycle Energy and Greenhouse Gas Emissions of Nuclear Energy: A Review. *Energy Conversion and Management*, 49 (8), August, 2178–2199.

Leonard A (2010) *The Story of Stuff*. London: Constable.

Leopold A (2001/1949) The Land Ethic. In M Zimmerman, J B Callicott, G Sessions, K Warren, & J Clark (eds.) *Environmental Philosophy*, 3rd edition. Upper Saddle River, NJ: Prentice Hall.

Lewis S & Maslin M (2015) Defining the Anthropocene. *Nature*, 519 (7542), 171–180.

Lima M (2011) *Visual Complexity: Mapping Patterns of Complexity*. New York, NY: Princeton Architectural Press.

Logan R (2014) Chapter 2: What Is Information? Why Is It Relativistic and What Is Its Relationship to Materiality, Meaning and Organization. Toronto: DEMO Publishing. Accessed online: http://demopublishing.com/ book/ what-is-information/chapter-2/

Logan R & Stokes L (2004) *Collaborate to Compete: Driving Profitability in the Knowledge Economy*. Toronto and New York: Wiley.

Lorde A (1984) The Master's Tools Will Never Dismantle the Master's House. In *Sister Outsider: Essays and Speeches* (2007). Berkeley, CA: Crossing Press, 110–114.

Louv R (2006) *Last Child in the Woods: Saving Our Children from Nature-Deficit Disorder*. London: Atlantic Books.

Lovins A, Lovins H, & Hawken P (2007/1999) *Natural Capitalism: Creating the Next Industrial Revolution*. London: Little Brown and Company.

Luciano D (2015) The Inhuman Anthropocene. *Avidly*, *Los Angeles Review of Books*. Accessed online: http://avidly.lareviewofbooks.org/2015/03/22/the-inhuman-anthropocene/

Luck G W, Chan M A, Eser U, Gomez-Baggethun E, Matzdorf B, Norton B, & Potschin M B (2012) Ethical Considerations in On-Ground Applications of the Ecosystem Services Concept. *BioScience*, 62 (12), 1020–1029.

Lukes S (1974/2005) *Power: A Radical View*, 2nd edition. Basingstoke, UK: Palgrave Macmillan.

Mackay D (2009) *Sustainable Energy Without the Hot Air*. Cambridge: UTI Cambridge Ltd.

Macy J (1995) Working Through Environmental Despair. In T Roszak, M Gomes, & A Kanner (eds.) *Ecopsychology*. San Francisco, CA: Sierra Club Books, 240–262.

Macy J (2009) The Great Turning. *Center for Ecoliteracy*. Accessed online: https://www.ecoliteracy.org /article/great-turning

Macy J & Johnstone C (2016) *Active Hope*. Novato, CA: New World Library.

Marshall G (2007) *Carbon Detox*. London: Gaia.

Marshall G (2009) The Psychological and Political Challenge of Facing Climate Change. Transformations, [Online] *Journal of Psychotherapists and Counsellors for Social Responsibility*. Spring 2009. Accessed online: www.pcsr.org.uk/doc/ Transformations_Spring_09_copy_.pdf

Masud L, Valsecchi F, Ciuccarelli P, Ricci D, & Caviglia G (2010) From Data to Knowledge Visualizations as Transformation Processes Within the Data-Information-Knowledge Continuum. 14th International Conference Information Visualisation.

Mauss M (1925) *The Gift*. Translated by I Cunnison. London: Cohen & West.

McAlister C & Wood J (2014) *The Potential of 3D Printing to Reduce the Environmental Impacts of Production*. ECEEE Industrial Summer Study Proceedings: Retool for a competitive and sustainable industry. Stockholm: European Council for an Energy Efficient Economy.

McCandless D (2010) A Hierarchy of Visual Understanding. Information is Beautiful.net. Accessed online: http://www.informationisbeautiful.net/2010/data-information-knowledge-wisdom

McDonough W & Braungart M (2002) *Cradle to Cradle: Remaking the Way We Make Things*. New York: North Point Press.

McIntosh A (2012) The Challenge of Radical Human Ecology to the Academy. In L William, R Toberts, & A McIntosh (eds.) *Radical Human Ecology: Intercultural and Indigenous Approaches*, London: Routledge, 31–56.

McLuhan M (2001/1964) *Understanding Media*. London: Routledge.

McLuhan M & Flore Q (2001/1967) *The Medium Is the Massage*. London: Penguin Books.

McMurtry J (1999) *The Cancer Stage of Capitalism*. London: Pluto Press.

Meadows D (1999) *Leverage Points: Places to Intervene in a System*. Hartland, VT: The Sustainability Institute.

Meadows D (2008) *Thinking in Systems*. Edited by D Wright. London: Earthscan.

Meadows D, Meadows D, Randers J, & Behrens W (1972) *Limits to Growth*. New York, NY: New American Library.

Means R (1980) [Speech] *For America to Live, Europe Must Die*. Black Hills International Survival Gathering in the Black Hills of South Dakota. July 1980.

Meng J, Liu J, Xu Y, Guan D, Liu Z, Huang Y, & Tao S (2016) Globalization and Pollution: Teleconnecting Local Primary $PM_{2.5}$ Emissions to Global Consumption. *Proceedings of the Royal Society A*, 472, 20160380.

Merchant C (2001/1980) The Death of Nature. In M Zimmerman, J B Callicott, G Sessions, K Warren, & J Clark (eds.) *Environmental Philosophy*, 3rd edition. Upper Saddle River, NJ: Prentice Hall.

Metzer R (1995) The Psychopathology of the Human-Nature Relationship. In T Roszak, M Gomes, & A Kanner (eds.) *Ecopsychology*. San Francisco, CA: Sierra Club Books, 55–67.

Milestone J (2007a) *Universities, Cities, Design and Development: An Anthropology of Aesthetic Experience*. Unpublished doctoral thesis, Department of Anthropology, Temple University.

Milestone J (2007b) Design as Power: Paul Virilio and the Governmentality of Design Expertise. *Culture, Theory and Critique*, 48 (2), 175–198.

Mitchell W J T (1994) *Picture Theory: Essays on Verbal and Visual Representation*. Chicago, IL: University of Chicago Press.

Monbiot G (2014) Can You Put a Price on the Beauty of the Natural World? *The Guardian*. 22 April. Retrieved from: www.theguardian.com/commentisfree/ 2014/apr/22/price-natural-world-agenda-ignores-destroys

Moore J W (2014) 'The Capitalocene. Part I: On the Nature & Origins of Our Ecological Crisis'. Accessed online: http://www.jasonwmoore.com/uploads/The_Capitalocene__Part_I__June_2014.pdf

Moore J W (2015) *Capitalism and the Web of Life*. London: Verso.

Morgan F (2006) *The Power of Community: How Cuba Survived Peak Oil*. [Film] The Community Solution.

Mosco V (2014) *To the Cloud: Big Data in a Turbulent World*. Boulder, CO: Paradigm.

Muraca B (2016) From Capitalist Accumulation to a Solidarity Economy. [keynote presentation] 5th International Degrowth Conference, Budapest, 31 August.

Nadal A (2012) *RIO+20: A Citizen's Background Document*. Mexico: Center for Economic Studies, El Colegio de Mexico.

National Social Marketing Centre (2015) What Is Social Marketing? Accessed online: http://www.thensmc.com/what-social-marketing

Nicholson-Cole S (2005) Representing Climate Change Futures: A Critique on the Use of Images for Visual Communication. *Computers, Environment and Urban Systems*, 29, 255–273.

O'Connor J (1998) *Natural Causes*. New York, NY: Guilford Press.

O'Connor T (1995) Therapy for a Dying Planet. In T Roszak, M Gomes, & A Kanner (eds.) *Ecopsychology*. San Francisco, CA: Sierra Club Book, 149–155.

OECD (2011) Divided We Stand: Why Inequality Keeps Rising, OECD Income Distribution and Poverty Database. Accessed online: www.oecd.org/els/social/inequality

OECD (2015) *In It Together: Why Less Inequality Benefits All*. OECD Publishing. Accessed online: http://www.oecd-ilibrary.org/ – www.oecd.org/els/social/inequality

Ogilvy Earth (2012) *From Greenwash to Great*. Accessed online: https://assets.ogilvy.com/truffles_email/ogilvyearth/Greenwash_Digital.pdf

Olsen J (1999) *Nature and Nationalism: Right-Wing Ecology and the Politics of Identity in Contemporary Germany*. New York: St. Martin's Press.

Onafuwa D, Bloom J, & Cole T (2016) Privileged Participation: Allying with Decoloniality in a Difficult Climate, [workshop] Carnegie Mellon University's School of Design.

Ong W J (1982) *Orality and Literacy*. London: Routledge.

Oreskes N (2010) *Merchants of Doubt: How a Handful of Scientists Obscured the Truth on Issues from Tobacco Smoke to Global Warming*. New York, NY: Bloomsbury Press.

Orr D (1992) *Ecological Literacy*. Albany, NY: State of New York Press.

Orr D (2002) *The Nature of Design*. Oxford: Oxford University Press.

Orr D (2004) *Earth in Mind*. London: Island Press.

Orr D W (2007) *The Designer's Challenge: Four Problems You Must Solve*. Commencement address, University of Pennsylvania, 14 May. Accessed online: https://www.ecoliteracy.org/article/designers-challenge-four-problems-you-must-solve

Orton D (1989) Sustainable Development or Perpetual Motion? Circles of Correspondence, *The New Catalyst*, Number 23, Spring 1989. Accessed online: http://home.ca.inter.net/~greenweb/Sustainable_Development.html

Ortony A (1979) *Metaphor and Thought*. Cambridge: Cambridge University Press.

Ostrom E (1990) *Governing the Commons*. Cambridge: Cambridge University Press.

Oxfam: Fuentes-Nieva R & Galasso N (2014) *Working for the Few. Political Capture and Economic Inequality*: Oxfam Briefing Paper. Oxford: Oxfam.

Oxfam: Hardoon D (2017) *An Economy for the 99%:* Oxfam Briefing Paper. Oxford: Oxfam International.

Packard V (1960) *The WasteMakers.* London: Longmans.

Papanek V (1999/1975) Edugraphology – The Myths of Design and the Design of Myths. In M Bierut, J Helfand, S Heller, & R Poynor (eds.) *Looking Closer 3: Classic Writings on Graphic Design*. New York: Allworth Press, 251–255.

Papanek V (2000/1971) *Design for the Real World: Human Ecology and Social Change*, 2nd edition. Chicago: Academy Chicago Publishers.

Parr A (2009) *Hijacking Sustainability*. London: MIT Press.

Paton B & Dorst K (2010) Briefing and Reframing at Design Thinking and Research Symposium Eight 2010 (DTRS8 2010) Sydney, 19–20 October 2010.

Peck J (2010) *Constructions of Neoliberal Reason*. Oxford: Oxford University Press.

Peck J (2013) Explaining (with) Neoliberalism. *Neoliberalism*, *Crisis and the World System*, University of York. 3 July.

Peters G P, Minx J C, Weber C L, & Edenhofer O (2011) Growth in Emission Transfers via International Trade from 1990 to 2008. *PNAS*, 108 (21), 8903–8908.

Piketty T (2014) *Capital in the Twenty-First Century*. Cambridge, MA: Harvard University Press

Pindar I & Sutton P (2000) Translators' Introduction in Guattari F, *The Three Ecologies*. London: Continuum.

Plumwood V (1999) Ecological Ethics from Rights to Recognition: Multiple Spheres of Justice for Humans, Animals and Nature. In N Low (ed.) *Global Ethics and Environment*. London: Routledge, 188–217.

Plumwood V (2002) *Environmental Culture: The ecological Crisis of Reason*. Oxon: Routledge.

Polanyi K (2001/1944) *The Great Transformation*. Boston, MA: Beacon Press.

Pope Francis (2015) *Encyclical Letter Laudato Si*, *On Care for Our Common Home*. Vatican. Accessed online: http://w2.vatican.va/content/francesco/en/encyclicals/documents/papa-francesco_20150524_enciclica-laudato-si.html

Posey D A (1999) *Cultural and Spiritual Values of Biodiversity*. United Nations Environment Programme. A Complementary Contribution to the Global Biodiversity Assessment. London: Intermediate Technology Publications.

Poynor R (2006) *Designing Pornotopia*. London: Lawrence King.

Purchase G (1997) *Anarchism and Ecology*. London: Black Rose Books.

Räsänen M & Nyce J (2013) The Raw Is Cooked: Data in Intelligence Practice. *Science Technology Human Values*, 38 (5), 655–677.

Ravetz J (2005) The Post-normal Sciences of Precaution', Conference on Uncertainty and Precaution in Environmental Management. *Water Science & Technology*, 52 (6), 11–17.

Reason P (1998) Political, Epistemological, Ecological and Spiritual Dimensions of Participation. *Studies in Cultures, Organizations and Societies*, 4, 147–167.

Reason P & Bradbury H (2006)*The SAGE Handbook of Action Research*. London: Sage.

Reich R (2015) Friction Is Now Between Global Financial Elite and the Rest of Us. *The Guardian*. 11 Nov. Accessed online: https://www.theguardian.com/commentisfree/2015/nov/11/us-uk-politics-economics

Resilience Alliance (2017) *Resilience*. Accessed online: http://www.resalliance.org/resilience

Rittel H W & Webber M M (1973) Dilemmas in a General Theory of Planning. *Policy Sciences*, 4 (2), 155–169.

Roberts D (2013, 17 April) None of the World's Top Industries Would Be Profitable If They Paid for the Natural Capital They Use. *Grist*. Accessed online: http://grist.org/ business-technology/none-of-the-worlds-top-industries-would-be-profitable-if-they-paid-for-the-natural-capital-they-use

Rockström J et al. (2009) Planetary Boundaries: Exploring the Safe Operating Space for Humanity. *Ecology and Society*, 14 (2), 32.

Rockström J et al. (2011) A Safe Operating Space for Humanity. *Nature*, 476 (282), August.

Roser M (2015a) [Tweet] On May 10th 2015 Roser tweeted: "Declining Racial Violence in the US since 1882" (@MaxCRoser)

Roser M (2015b) Treatment of Minorities – Violence and Tolerance. OurWorldInData.org. Retrieved from: goo.gl/68xybg

Roser M (2015c) Natural Catastrophes. OurWorldInData.org. Retrieved from: goo.gl/EwlrjV

Roszak T (1992) *The Voice of the Earth*. London: Bantam Books.

Roszak T (1995) Where Psyche Meets Gaia. In T Roszak, M Gomes, & A Kanner (eds.) *Ecopsychology*. San Francisco, CA: Sierra Club Books, 1–20.

Roszak T, Gomes M, & Kanner A (eds.) (1995) *Ecopsychology*. San Francisco, CA: Sierra Club.

RSA Action and Research Centre (2013) *The Great Recovery: Redesigning the Future*. Report 01: June 2013. London: RSA.

Rushkoff D (1996) *Media Virus*. London: Ballantine Books.

Rushkoff D (2016) *Throwing Rocks at the Google Bus: How Growth Became the Enemy of Prosperity*. London: Portfolio: Penguin.

Sachs W (1999) *Planet Dialectics*. London: Zed Books.

Sachs W (2010) *The Development Dictionary*. London: Zed Books.

Sacks D (2008) Green Guru Gone Wrong: William McDonough. *Fast Company*. 11 January. Accessed online: https://www.fastcompany.com/1042475/green-guru-gone-wrong-william-mcdonough

Santos B S (2007) *Cognitive Justice in a Global World*. Plymouth: Lexington Books.

Schling H (2011) Corporate Engagement at Hopenhagen. *Corporate Watch*, 49. London: Corporate Watch.

Schumacher E F (2010/1973) *Small Is Beautiful: A Study of Economics as if People Mattered*. London: HarperCollins.

Schwartz S H (1992) Universals in the Content and Structure of Values: Theoretical Advances and Empirical Tests in 20 Countries. In M P Zanna (ed.) *Advances in Experimental Social Psychology*, 25. Orlando, FL: Academic Press, 1–65.

Sears P (1964) Ecology – A Subversive Science. *Bioscience*, 14 (7), 11–13.
Sevaldson B (2013) Systems Oriented Design: The Emergence and Development of a Designerly Approach to Address Complexity. DRS/Cumulus Oslo 2013: 2nd International Conference for Design Research Education Researchers. Oslo, 14–17 May.
Sevaldson B (2016) A Library of Systemic Relations. SOD. Accessed online: http://www.systemsorienteddesign.net/index.php/giga-mapping/types-of-systemic-relations
Sewall L (1995) The Skills of Ecological Perception. In T Roszak, M Gomes, & A Kanner (eds.) *Ecopsychology*. San Francisco, CA: Sierra Club Books, 201–215.
Sewall L (1999) *Sight and Sensibility: The Ecopsychology of Perception*. New York, NY: Putnam.
Sewall L (2012) Beauty and the Brain. In P H Kahn Jr & P H Hasbach (eds.) *Ecopsychology: Science, Totems and the Technological Species*. London: MIT Press, 241–264.
Share the World's Resources (STWR) (2014) *A Primer on Global Economic Sharing*. London: Share The World's Resources. Accessed online: http://www.sharing.org/information-centre/reports/primer-global-economic-sharing
Shellenberger M, Nordhaus T, Brand S, Lynas M, Pielke R Jr, Sagoff M, Ellis E, & others (2015) *An Ecomodernist Manifesto*. Accessed online: http://www.ecomodernism.org/
Sheppard P (1995) Nature and Madness. In T Roszak, M Gomes, & A Kanner (eds.) *Ecopsychology*. San Francisco, CA: Sierra Club Books, 21–20.
Shiva V (1988) Reductionist Science as Epistemological Violence. In A Nandy (ed.) *Science, Hegemony and Violence*. Oxford: Oxford University Press.
Shiva V (2005) *Earth Democracy*. London: Zed Books.
Simms A, Johnson V, & Chowla P (2010) *Growth Isn't Possible*. London: New Economics Foundation.
Smith R (2007) [Keynote Lecture] Carpe Diem: The Dangers of Risk Aversion, *Lloyd's Register Educational Trust Lecture*. 29 May 2007. London.
Solon P (2014, August 20) We Live CAPITALCENE. Not 'Anthropocene' Because the Logic of Capital Drives Disruption of Earth System. Not Humans in General. #ESU2014. [tweet] Accessed online: https://twitter.com/pablosolon/status/502037442556596224
SourceWatch (2011) BP and Greenwashing. Accessed online: http://www.sourcewatch.org/index.php?title=BP_and_Greenwashing
Spash C (2016) Science and Uncommon Thinking [keynote presentation] 5th International Degrowth Conference, Budapest, 31 August.
Specter M (2014) Seeds of Doubt: An Activist's Controversial Crusade Against Genetically Modified Crops. *The New Yorker*. 25 August.
Spretnak C (1997) *The Resurgence of the Real*. New York, NY: Addison-Wesley.
Steffen W, Crutzen P J, & McNweill J R (2007) The Anthropocene: Are Humans Now Overwhelming the Great Forces of Nature? *Royal Swedish Academy of Sciences*, 35 (8).
Steffen W et al. (2015a) Planetary Boundaries: Guiding Human Development on a Changing Planet. *Science*, 347 (6223), 736.
Steffen W et al. (2015b) The Trajectory of the Anthropocene: The Great Acceleration. *The Anthropocene Review*, 2 (1), 81–98.
Sterling S (1993) Environmental Education and Sustainability: A View from Holistic Ethics. In J Fien (ed.) *Environmental Education – A Pathway to Sustainability*. Geelong: Deakin University.
Sterling S (2001) *Sustainable Education—Re-visioning Learning and Change*, Schumacher Briefing no. 6. Dartington: Schumacher Society/Green Books.
Sterling S (2003) *Whole Systems Thinking as a Basis for Paradigm Change in Education*. PhD, University of Bath.
Sterling S (2011) Transformative Learning and Sustainability: Sketching the conceptual ground. *Learning and Teaching in Higher Education*, 5, 17–33.

Stern N (2007) *The Economics of Climate Change – The Stern Review*. Cambridge, UK: Cambridge University Press.

Stiglitz J (2001) Foreword in K Polanyi [1944] *The Great Transformation*. Boston, MA: Beacon Press.

Stiglitz J (2013) *The Price of Inequality*. London: W W Norton & Company.

Stockholm Resilience Centre (2017) The Nine Planetary Boundaries. [website] Accessed online: http://www.stockholmresilience.org/research/planetary-boundaries/planetary-boundaries/about-the-research/the-nine-planetary-boundaries.html

Stone M K (2015) *Applying Ecological Principles*. Centre for Ecological Literacy. [website] Accessed online: http://www.ecoliteracy.org/nature-our-teacher/ecological-principles

Strega S (2005) The View from the Poststructuralist Margins: Epistemology and Methodology Reconsidered. In L Brown & S Strega (eds.) *Research as Resistance*. Toronto: Canadian Scholars Press.

Sullivan S (2013) *Financialisation, Biodiversity Conservation and Equity: Some Currents & Concerns*. Penang: Third World Network.

Swaine J & Laughland O (2015) Eric Garner and Tamir Rice among Those Missing from FBI Record of Police Killings. *The Guardian*. 15 October 2015. Accessed online: www.theguardian.com /us-news/2015/oct/15/fbi-record-police-killings-tamir-rice-eric-garner

Thackara J (2015a) *How to Thrive in the Next Economy*. London: Thames & Hudson.

Thackara J (2015b) Bioregions: Notes on a Design Agenda. John Thackara. 23 April 2015. Accessed online: http://thackara.com/place-bioregion/bioregions-notes-on-a-design-agenda/

Tonkinwise C (2007) Positions/Practicing Sustainability by Design: Global Warming Politics in a Post-awareness World. In J Merwood-Salisbury (ed.) *Scapes*. New York, NY: Parsons The New School for Design, 4–13.

Townsend S & Crompton T (2010) Are People Fundamentally Selfish and Self-motivated? *The Ecologist* [website]. 10 March 2010. Accessed online: http://tiny.cc/um7jiy

Tufte E (1990) *Envisioning Information*. Graphics Press, Cheshire: Graphics Press.

Tufte E (1997) *Visual Explanations: Images and Quantities, Evidence and Narrative*. Cheshire, CT: Graphics Press.

Tufte E (2010) Workshop with Edward Tufte. Denver: Denver Marriott Tech Center, 11 June.

Turner F (1994) The Invented Landscape. In A Baldwin, D Dwight, J Petschen, & C Pletsch (eds.) *Beyond Preservation: Restoring and Inventing Landscapes*. Minneapolis, MN: University of Minnesota Press, 34–66.

Ulrich W (1998) Systems Thinking as if People Mattered: Critical Systems Thinking for Citizens and Managers. Working Paper No. 23, Lincoln School of Management, UK.

United Nations Environment Programme (2011) *Towards a Green Economy: Pathways to Sustainable Development and Poverty Eradication*. New York: UNEP.

United Nations Food and Agriculture Organization [UN-FAO] (2017) FAO: Food and Agriculture Organization of the United Nations. Accessed online: http://www.un.org /youthenvoy/2013/09/fao-food-and-agriculture-organization-of-the-united-nations/

Valsecchi F, Ciuccarelli P, Ricci D, & Caviglia G (2010) The Density Design Lab: Communication Design Experiments among Complexity and Sustainability. In *Proceedings Cumulus Conference 2010 Young Creators for Better City and Better Life*, Shanghai, China, September 2010.

Varoufakis Y (2012) Why Valve? Or, What Do We Need Corporations for and How Does Valve's Management Structure Fit into Today's Corporate World? *Value Economics: A Blog by Yanis Varoufakis*. 3 August 2012. Accessed online: http://tiny.cc/iucxiy

Vatn A (2000) The Environment as a Commodity. *Environmental Values*, 9, 493–509.

Vatn A (2010) An Institutional Analysis of Payments for Environmental Services. *Ecological Economics*, 69, 1245–1252.

Venturini T et al. (2014) *Climaps by EMAPS in 2 Pages* (A Summary for Policymakers and Busy People in General). Accessed online: http://ssrn.com/abstract =2532946

Venturin T et al. (2015) Designing Controversies and Their Publics. *Design Issues*, 31 (3), 74–87.

Wahl D (2006) *Design for Human and Planetary Health*. Unpublished doctoral thesis, Dundee University.

Wahl D C (2016) *Designing Regenerative Culture*. Axminster, UK: Triarchy Press.

Warren K (1999) Care-Sensitive Ethics and Situated Universalism. In N Low (ed.) *Global Ethics and Environmentalism*. London: Routledge.

Warren K (2001/1990) The Power and Promise of Ecological Feminism. In M Zimmerman, J B Callicott, G Sessions, K Warren, & J Clark (eds.) *Environmental Philosophy*, 3rd edition. Upper Saddle Rive, NJ: Prentice Hall, 322–342.

Weheliye A (2014) *Habeas Viscus*. Durham, NC: Duke University Press.

Welsh I (2010) Climate Change: Complexity and Collaboration Between the Sciences. In C Lever-Tracy (ed.) *Routledge Handbook of Climate Change and Society*. London: Routledge

Wheatley M (2006) *Leadership and the New Science: Discovering Order in a Chaotic World*. Oakland, CA: Berrett-Koehler Publishers.

White D, Rudy A, & Gareau B (2015) *Environments, Natures and Social Theory: Towards a Critical Hybridity*. London: Palgrave.

Whitehead A N (1925) *Science and the Modern World*, Vol. 55 of the Great Books of the Western World Series. New York, NY: Macmillan Company.

Whitmarsh L, O'Neill S, Seyfang G, & Lorenzoni I (2009) Carbon Capability. In A Stibbe (ed.) *The Handbook of Sustainability Literacy*. Dartington: Green Books, 124–129.

Wijkman A & Rockstrom J (2011) *Bankrupting Nature: Denying Our Planetary Boundaries*. Oxon: Earthscan/Routledge.

Wilkinson R & Pickett K (2010) *The Spirit Level: Why Equality Is Better for Everyone*. London: Penguin.

Willis A M (2010) Can Complexity Be Contained? *Design and Complexity Conference*, University of Montreal, 7–9 July 2010. Montreal: Design Research Society.

Wilson E O (1984) *Biophilia*. Cambridge: Harvard University Press.

Windle P (1995) The Ecology of Grief. In T Roszak, M Gomes, & A Kanner (eds.) *Ecopsychology*. San Francisco, CA: Sierra Club Books, 136–148.

World Food Program [WFP] (2017) Frequently Asked Questions (FAQs). Accessed online: http://www.wfp.org/hunger/faqs

WWF-INT (2010) *The Living Planet Report 2010*. Gland, Switzerland: WWF-International.

WWF-INT (2012) *Living Planet Report 2012*. Gland, Switzerland: WWF-International.

WWF-INT (2014) *Living Planet Report 2014*. Gland, Switzerland: WWF-International.

WWF-UK (2010) *Annual Report 2010*. Godalming, UK: WWF-UK.

Zalasiewicz J et al. (2015) When Did the Anthropocene Begin? A Mid-twentieth Century Boundary Level Is Stratigraphically Optimal. *Quaternary International*, 383, 204–207.

Index

Abrams, David 116–18
acceleration 9, 38, 40, 75, 142, 160
acknowledgement 24, 74, 135–8
actionable 175, 181
activists 25–6, 30–7, 52, 64–9, 72, 83, 105, 109–11, 145–50, 169, 173
    design activism (*see* design)
    documented deaths (of) 36
advertising 29, 46, 134, 135
aesthetics/aestheticize 3, 16, 21, 30, 53, 62, 117–18, 121, 130–1
agency 21, 32, 58, 59, 63, 83, 151, 158, 175–6, 182
agroecology/agriculture 108, 164
allyship 26
alternatives 111
anarchism 105–7, 108
animate 109, 130
anthropocene 8–11, 164
anti-ecological 118
anti-fascist 106
appropriation 76, 106, 151, 183
Armstrong, Rachel 11, 167
arts, the 56, 82
austerity 33, 34, 76, 105, 111, 158
authoritarianism 17–18, 30, 32–3, 105, 108, 111, 160, 162, 171, 175

backgrounding 58, 72
Bacon, Frances 51–2, 118
Ban, Ki-moon 146
bandwidth 116
Barry, Marie-Ann 114, 116, 119
Bateson, Gregory 2, 3, 62
    levels of learning and communication 77, 85
    and mental health 141
    patterns that connect 84, 115
    perception 116
    soft systems approach 61, 85–6
    Steps to an Ecology of Mind 2–3, 62–4, 119, 141
Bateson, Mary Catherine 118, 119

Battiste, Marie 68
Bennett, Jane 67
Benyus, Janine 88
Berger, John 1
Bernay, Edward 21
Bertalanfry, Ludwig von 84
beyond petroleum 147
bias 52, 59, 80, 86, 92, 117, 135, 171
Biko, Steven 33
biocentrism 104
biodiversity crisis 152
biomimicry 88
biophilia 104, 139
biopiracy 68
bioregion 89
Black Lives Matter 173
Bookchin, Murray 104, 107
Bourdieu, Pierre 27–32
Breakthrough Institute 164
Buckminster Fuller 15, 61
Bullard, Robert 103
business-as-usual 7, 10, 30, 32, 46, 77, 149, 185

capital accumulation 103
capitalism
    and design 16–19, 31, 37
    flows of information 38
    and growth 44
    ideology 31, 106, 162
    logic, invisible hand, etc. 26, 158
    Marxist critique 107
    second contradiction 43, 108
    and value systems 134, 182
capitalocene 10
Capra, Fritjof 3, 44, 82, 88
carbon capability 91
carbon intensity 95
care 2, 20, 60, 65, 71, 72, 75, 136, 152, 157, 165, 172
causality 125–7
Center for Ecological Literacy 81–2, 88

Centre for Alternative Technologies (CAT) 92
Chapman, Jonathan 21
China, carbon emissions 94–6
Chomsky, Noam 37
circular economy 89–90, 161, 166–7
Climate Camp 36–7, 111, 147–8
climate change 6–10, 38, 40, 43, 45, 59, 89, 91–5, 100, 103, 108–10, 145–51, 153, 156, 158, 160, 164, 174–5, 176–81, 192
   climate contrarian 179–80
   climate discourses 177–80
cognitive capacities 1, 19, 114, 116
cognitive dissonance 135, 137–8
cognitive frames 118
Cohen, Stanley 135–7
Coke 147–9
colonialism/colonization (or decolonizing) 28, 64, 66–70, 72, 111, 162, 182
common cause 133, 157–8
commonality 44, 67, 168–9
Commoner, Barry 107
commons, the 48, 55, 101, 104, 106, 150, 168
communication
   campaigns (vs. marketing) 145–51
   explicit/implicit and contradictory 178 (*see also* neoliberalism)
   levels and processes 77–80, 171, 175–6
   revolutions in modes 104, 113–18
communist 108, 111
community-based 105, 168
comparisons 48, 121, 123, 125, 130, 171–2
complex adaptive systems 60, 84, 95
complexity 5, 58–60, 113, 170, 176–7
   cognition 18
   navigation 22–3, 25, 176–7, 184
   negating 25, 40, 53, 172
   obscured/reduced/abstracted 52, 173
   visual 18, 128
   as worldview 55, 182–4
conflict of interest 146, 150
Connolly, William 17, 31
consequences 169
constituencies 174, 180
constructed scarcities 76, 92
consumer 5, 16–17, 19, 21, 30, 46, 73, 74, 76–7, 93, 104, 106, 134–6, 138, 145–6, 148–51, 160, 161, 165, 182
consumption 7, 19, 21, 31, 44–5, 51, 59, 76–7, 89, 146, 164, 167
consumption (conspicuous) 16, 19, 29, 48, 77, 114, 134
context 123–5

contradiction 138, 153–9, 153, 177–81
control 52, 85
controversy 145, 150, 160, 170, 176–81
cooperative 106–7, 111, 133, 184
co-opting 150
COP-15 (Copenhagen) 110, 145–51
COP-21 (Paris) 142
corporate social responsibility 75, 145, 148
counterpower 32–7
Cox, Robert 47
cradle-to-cradle 166
crisis
   as catalyst 183–5
   discipline 47
   of perception 130
critical consciousness 33, 82
Cuba 112
Curtis, Adam 61, 86
cybernetics 84, 85
cycles 89–90

Daly, Herman 161
data 175–6
dark data 174–96
   data visualization 170–81
   datawash 31, 37, 174–5
Davis, Angela 65
De Sousa Santos, Boaventura 53
death of nature 118
decolonize/decolonizing 28, 64, 66–70, 72, 111, 182
decolonizing design 72
decoupling 43, 93, 164
deep ecology 104–5
deep horizon oil spill 147
defence mechanisms 138, 141
degrowth. *See* growth
Deloria, Vine 68
democratic/democracy 19, 32–3, 44, 57–8, 95, 105–8, 149–50, 152, 158, 169
denial 56, 135–8, 183
   climate change 178–81
   stages and types 135–6
Density Design 124
depth 117
Deregowski, Jan B. 118
design, types of
   service 22–3
   social innovation 23–4
   transition 25
   speculative 24
   activism/activist 26
   system oriented design 98, 121

Design Council UK 22
designerly 115, 123
design intelligence. *See* intelligence, design
desire/desireable 1, 5, 15, 16, 29, 30, 47, 83, 86, 107, 121, 131, 132, 182, 186, 188
devitalized 52
digital positivism 172, 181
D'Ignazio, Catherine 170, 175
direct action 36, 104, 111, 147–8
directed selective attention 116
disciplinary discourse 28, 30
discourse
    climate contrarian 159, 162
    definition 29
    environmental 106–7, 162, 177–81, 182
discursive confusion 177–81
disengagement 58–60, 165–6
disrupt/disruptive 7, 10, 72, 74, 82, 85, 87, 93, 111, 166, 184
domains of seeing, thinking, doing 79
domination 21, 52, 64–6, 104, 107, 110, 119, 139, 182
    logic of 65, 110, 118
    modes of 27, 31, 64, 107, 118
    over nature 52, 86
Dondis, Donis 114
Drysek, John 29
DSM-5, 134
dualist 51, 58
Dunne, Anthony 24
dynamic balance 88, 98–101
dysfunction 28, 51, 62–3, 71, 82, 84, 97–8, 106, 115, 139–40, 167

ecoanxiety 139
ecocene 11, 182–4
ecocide 138
ecoeffectiveness 166
eco-fascism 61
ecofeminism 28, 102, 104, 110–11
ecoism/ecoist 28–31, 37, 162, 184
ecological
    ethics 2–3, 8, 20, 25, 47, 61, 64, 67, 71–2, 105, 140, 184
    identity 130–42
    learning 74–87, 182–3
    thought 2, 19, 51, 53, 55, 61, 62, 65, 74, 83, 87, 104, 117, 184
    troubling 67
ecological assessment tools 61, 98–100
ecological economics 93, 154–5
ecological footprint 7, 48, 61, 78, 88, 98–9, 111, 129–30, 138

ecological literacy 1, 5, 16, 30, 74–87, 88, 97, 115–16, 182
    definition 74–7, 97, 102, 182
    and design 15, 25, 86–7
    typologies of (mode one/mode two) 80–2, 104
ecologically restorative 109
ecological perception 1–2, 113–31
ecology 2, 5, 47, 53–5, 61, 62–4, 67, 74, 82, 83, 87, 142, 152, 182
    and images 53
ecomodernism/ecomodernist 104, 164–5, 169, 178, 185
economics 5, 7, 10–11, 17, 19, 31, 33–4, 38–45, 47–8, 59, 71–2, 89–93, 98–100, 102–8, 145, 152, 154–5, 161–4, 167
ecoparalysis 139
ecopedagogy 80–2, 83
ecophobia 139
ecopsychology 70, 138–42
ecosystem services 44, 81, 153, 154, 156, 158
embedded 117, 118, 139, 141
embodied 28, 55, 81, 131, 141
embodied energy 88, 92, 94–5, 168
emergence 60, 95–8
emergent order 17–19
empathetic/empathy 64–5, 130–1, 133–4, 157
empirical 51, 53, 57, 62, 125, 137, 175
energy 91–5
    embodied 94–5 (*see also* embodied energy)
energy descent 89, 95, 105, 112
energy literacy 91–5
Energy Return On Investment (EROI) 92–3
environmental
    justice 102–4, 110, 160, 162–3
    racism 103
    skepticism 162
epistemic selectivities 65
epistemological blindness 64, 66
epistemological error 62–73, 74, 116, 118, 139, 156
    addressing by design 123–31
epistemological flexibility 57, 85, 117
epistemological privilege 65–6
epistemological violence 52
ethic/ethics (ecological) 2–3, 8, 20, 25, 47, 61, 64, 67, 71, 105, 140, 184
Ewen, Stuart 21
exergy 93
exploitation 17, 30, 31, 45, 46, 51, 52, 58–60, 64, 72, 74, 108, 119, 145, 156, 158
externalities 47, 152–5
extinction 56, 88, 100, 102, 156, 184

facts/factual information 37, 57, 74, 134–6, 141, 171, 174–5
false consciousness 32
feedback metaphors 154
feminist
    black feminism 67–8
    data visualization 170–5
    ecofeminist 52, 58, 64–5, 72, 111
    intersectional 64, 72, 110
    movement, the feminist 72, 111, 118
    principles and progress 111
    standpoint theory 66, 170
financialization 152–9
First Things First Manifesto 19
flying 108
food security 108
food sovereignty 108
Foster, John Bellamy 107
Foucault, Michael 28–32
fragmentation 116, 119
framing 118–19, 157, 182
freedom 17, 61, 69, 111, 162
Freud, Sigmund 139
Friedman, Milton 183
Friere, Paulo 83
Fry, Tony 25, 78
Fuad-Luke, Alastair 26

Garland, Ken 19
GDP 44, 48, 156
generative/regenerative 7, 41, 43–5, 52, 53, 68, 72, 118, 169, 184
geographic maps. *See* maps, geographic
Giga-mapping 85, 122–3
Goleman, Daniel 65, 73
Graeber, David 105, 146
great turning, the 142, 184
Greece 113
green economy 110, 152–9
Greenpeace 46, 111, 124
greenwash 31, 37, 77, 145, 147, 150, 151
growth 39–40, 106, 164
    degrowth 40, 44–5, 48, 76
    and ecological design 89
    Prosperity Without Growth? 43
    types of, 40–5, 106
Guattari, Felix 2–3, 17, 182, 184

Haeckel, Ernest 53, 100
Hall, Stuart 171
Haraway, Donna 10, 65
Harding, Sandra 66
Harding, Stephen 55

hierarchy 17, 27, 30, 68, 104, 105, 107, 121, 158, 178
Horn, Robert 18, 115, 118, 125, 127
Humantific 22

identity 80, 132–5, 157
ideology/ideological 26, 28, 29, 31–2, 77, 105, 117, 159, 161, 170, 174, 177, 178, 180
indigeneity 66–70
Indigenous Environmental Network 109
indigenuity 209–10
industria 162, 165–7
inequality 37, 162–4
information deficit model 134
information theory 175–8
infrastructure 92
innerism 136
innovation 5, 16, 22–3, 38, 57, 88, 147, 160–2, 164–6
instrumental 85, 86
instrumentalization 59, 67, 154
intelligence
    design 115
    ecological 73
    emotional 65
    visual 113–15
interests 1, 2, 10, 17, 18, 21, 26, 28–30, 40–1, 45, 56, 58–9, 63–7, 77, 87, 103, 105, 114, 119, 125, 135–6, 145–51, 155, 158, 160, 164–5, 170–80, 182–3
international advertising association 146
Irwin, Terry 22–5

Jackson, Tim 43, 93
Jameson, Fredric 24
Jevons Paradox 93
just sustainability 76

K-cup 161–2, 169
Kahan, Dan 134–5
Kallis, Giorgos 164–5
Kari-Oca 2 Declaration 110
Kimmerer, Robin Wall 70
kin, kinship 69, 72
Klein, Naomi 10, 183
knowledge
    and capitalism 38, 40, 46, 151
    definition 175–6
knowledge visualization 121, 175–81
Korzybski, Alfred 56, 115
Kossoff, Gideon 25
Kuhn, Thomas 56–7

La Duke, Winona 109
La Via Campesina 109
Laboratory of Insurrectionary Imagination 142
laissez-faire 41, 106, 162
Lakoff, George 134
land ethic, the 71
Latour, Bruno 10, 67, 164
Leopold, Aldo 71
levels of learning and communication 77–8
leverage points in a system 84
life cycle analysis 94, 167
Lima, Manuel 127–8
Living Planet Index 7
Lorde, Audre 162
Luddites/Luddite Rebellion 168
Lukes, Stephen 29–30

MacDonough, William 166
Mackay, David 92
Macy, Joanna 140, 142
maker, cultures DIY 168
Malthus, Thomas 103
map/mapping 56–7, 79, 85, 94, 113, 115, 122–8, 130, 176–81
    conceptual 115, 125
    controversies 176–81
    geographic 100, 125
    giga-mapping 85, 122–3
    mapping unknowns 97, 127
    mental maps 118
    systems 85
map, is not the territory 56, 115
market, the 25, 38
    'free' myth 41
    logic of 152
marketing 134, 135
Marxism/Marxist 69, 107–8
master's house 162
material economy 175
Mauss, Marcel 146
McLuhan, Marshall 18, 113
Meadows, Donella 84
meaning/meaningful 118, 130, 170, 175–6
Means, Russell 52, 69
Merchant, Carolyn 52, 65, 118–19
metaphors 118–19
Milestone, Juris 30
Miller, George 116
modernist, modernity, modernization 51–2, 55–8, 60, 63–4, 67–71, 74, 93, 104, 169
    ecological modernization (see ecomodernism)
    ultra-modernism/post-modernity 51, 55–7, 69

Moore, Jason 10
Mosco, Vincent 172
motivation 156–8
mutual aid 26, 106–7, 111
mutualistic 118–19
myopic/myopically 76, 130

Naess, Arne 104
narcissism/narcissistic 130, 133, 134
natural capital 153–8
natural catastrophes 173–4
nature's patterns and processes 88
neoliberal/neoliberalism
    assumptions 30–1, 37, 102, 154
    definition 5–6, 17–18, 29, 31, 158–9
    and design 5, 30
    discourse 178–80
    discursive confusion 159, 177–81
    policy 111, 158, 163
    subject and sensibilities 5, 6, 17, 31, 102, 132
    and the environment 154, 158
nested systems 70–1, 85, 98, 100
network science 88
network visualization 113, 115, 127–8
networks 10, 18, 19, 28, 40, 72, 84, 86, 88–9, 99, 100–1, 105, 128, 132, 162
neutrality 5, 28–9, 59, 80, 84, 149, 158, 160, 162, 170, 175
Nordhaus, Ted 164
normalization 136–7
norms 134, 158, 167, 171, 187
nuclear power 92–3, 162, 164–5

obfuscations 159
objectification 59, 67, 70
objective/objectivity 51, 63, 66, 170–1, 180
observing, observer 52–3, 85, 117, 131, 171–2
O'Connor, James 43
Occupy Design 26, 34–5
Oglivy Earth 147, 151
Ong, Walter 114
oral cultures 103–4
Orr, David 3, 15, 60–1, 74–5, 86–7, 183
Ostrom, Elinor 146
othering 31, 37, 59, 64–5
overpopulation 103, 106

Packard, Vince 20
Papanek, Victor 20
paradigm 11, 22–3, 44, 55, 56–9, 78–9, 84, 89, 117, 184

Index 205

participant/participation/participatory 22, 26, 52, 57, 68, 79, 86, 88, 119, 131, 138, 150, 158, 182
pathology 137, 139, 140
pattern 17, 21, 28, 48, 79, 83–4, 88–101, 103, 113, 115–16, 119–20, 121, 123, 128, 130–1, 141, 182
peak oil 92, 105, 112
perception 1–2, 28–9, 51–2, 60–3, 65, 75, 78–9, 81, 85, 113–31, 180, 184
perceptual habits 116, 117–20, 130–1, 180
permaculture 102, 104–5, 119
photography 113
pictorial superiority effect 116
planetary boundaries 6–7, 100
planned obsolescence 20, 160
pleasure 16, 130–1
Plumwood, Val 58–61, 65, 72, 162
Polanski, Karl 41
Poor Man's Guardian, The 34–5
Pope Francis 45
positivist/positivism (science) 51–5, 57, 60, 161, 172–3
post-environmentalists 164–5
post-modernity. *See* modernity
post-normal science 25, 57–8
power, theory of 29–30
Poynor, Rick 165
precautionary principle 88, 127
priorities 1–2, 15, 17–19, 30, 38–48, 74–5, 82, 84, 108, 119, 145, 148–9, 152, 160, 167, 181, 184
pro bono 148, 151
profit-seeking/-making/-able 18, 38, 47, 48, 107–8, 167, 182
progress 20, 44, 48, 62, 69, 93–4, 147, 150, 160–2, 166–7, 169, 173, 184
prosperity 11, 29, 39, 41, 43–4, 162, 184
protest 17, 36, 95, 111, 146–9, 150
psychic numbing 141
psychological 21, 25, 30, 61, 104, 133–42, 156, 184
psychoterratic states 139
Purchase, Graham 106

qualitative (mode/focus) 43–4, 59, 80, 85, 107, 117–18, 121, 123, 130–1, 159, 171–2, 176–7, 184
qualitative growth (economic) 37, 40–1, 43–4, 107
quantification bias/quantitative reasoning 51–2, 58, 60, 85, 130–1, 157–9, 161, 171–2
quantitative comparisons/methods 171
quantum physics 52, 117

Raby, Fiona 24
racism/racist 66–8, 172–4
reason, crisis of 58
reasoning (moral) 65. *See also* ecological, ethics
rebound effect 93
recycle 166
reductionism/reductionist 51–2, 53, 55–7, 60, 69, 79, 83, 95, 98, 116, 130, 142, 154, 156, 170–2
reductionist science 95
reflexivity 85, 86
regenerative 169. *See also* generative
regime of truth 28
regional (and bioregion) 81, 89, 105–10, 125
regionalism 81, 106
relational perception 85, 113–31. *See also* ecological perception
relationality 122
relationships, types of 1, 5–6, 18, 25, 28, 52, 58, 63, 68–70, 72, 74–5, 77, 79, 83–6, 88, 95, 98, 109, 113–27, 131–3, 139–41, 154, 156, 158, 176–8
remote-controlling 17
remoteness (spatial, temporal, and technological) 30, 59, 71, 123–7
renewable energy 5, 44, 92, 105, 168
representation 27, 46–7, 115, 122, 127, 173, 177, 180, 182
repression 41, 139–40
resilience 6, 8, 44, 60, 82, 84, 87, 88–9, 100, 105, 108
responsibilization 21, 93
restorative 109, 131. *See also* regenerative and generative
rhetoric 153, 158, 159, 179–80
Rio+20 (2012) 110, 152
Rio Earth Summit (1992) 153
risk 3, 7–8, 30, 39, 41, 44, 46, 91, 100, 127, 140, 152, 156, 172, 174, 180–1
Role of Human Identity 133
Roser, Max 172–4
Roszak, Theodore 138
Royal Society of Arts 90

scarcity 43, 76, 90–3, 105–7
Schumacher, E.F. 153, 159
Schumacher College, 81–2
science/scientific 6, 8, 10, 25, 29, 45, 47, 52–8, 60, 62, 68, 74, 80, 85, 87, 88, 95, 121, 125, 128, 151, 160–2, 172, 175–6, 178–9
  post-normal science 25, 57–8
Science and Technology Studies (STS) 176
scientific revolution 51–3, 119

Seattle, Battle of 94–5
seeing (ways of) 1, 23, 79, 113, 116, 118, 120, 130–1
self-organization 15, 18, 84, 89, 95, 97, 106, 130
self-renewal or self-regulate 41, 75, 97, 98
sense-making 48, 118, 134
sensibilities 3, 5, 16–17, 29–31, 56, 104, 118, 130–2, 134, 175, 180, 185
Sewall, Laura 3, 116–17, 120, 130–1
Shellenberger, Michael 164
Shiva, Vandana 42, 52, 66–8
situated knowledge 65
Smuts, Jan 53, 61
social change 150
social ecology 107
social marketing 145–51
social movements 1, 15, 26, 32–7, 55, 56, 83, 102–11, 118, 136, 142, 145–51, 152, 158, 161, 168, 182–3
social relations 18, 80, 160
solidarity 26, 37, 72, 83, 89, 104, 139
soliphilia 139
soul 52, 70, 109, 185
spirit/spirituality 16, 69–70, 104, 109, 119, 138, 153, 185
Stahel, Walter R. 166
States of Denial 135–7
status quo 5, 30, 60–1, 76, 82, 145, 166
Sterling, Stephen 63–4, 77–8
Stern, Lord Nicholas and the *Stern Report* 108
structures/system structures 76, 87, 93, 102, 106, 166, 166, 167, 184
subjective condition 138
subjective grip 17
subjectivity/subjectivities 1–3, 5, 17, 29–30, 82, 85, 87–8, 141, 172, 182
substitutability 154–6, 164
sustainability 75, 100–1, 102, 146, 160, 167, 169, 182, 183, 185
Sustainable Energy – Without the Hot Air 92
sustainment 76
symbolic violence 27–32
systems (leverage points) 84
systems oriented design 98, 121–3
systems thinking 82, 83–6
  hard and soft 85
  and images 85

tacit knowledge 18
technocracy 169
technocratic/technocrat 93, 141, 164, 169
technofix 160–9, 181
technology 160–9

technosphere 166
Thackara, John 89, 93, 184
theory 21
3-D printing 167–8
three parts of paradigm 78, 79
threshold 102, 127, 156
TINA 'there is no alternative' 37
Tonkinwise, Cameron 25, 93
traditional ecological knowledge 57, 68–70, 102, 109–10
transformation design 22
transformative learning 78–82
transition design 25
transition towns/movement 25, 89, 105, 111
Tufte, Edward 15, 125, 180
uncertainty 92, 100, 127, 175

UNFCCC 7, 177
unintended consequences 71, 161, 168, 169, 174, 181
unknown territory maps 97, 127

values 21, 31, 111, 145
  and capitalism 38, 146
  and communication 154, 157
  feminist 71, 111
  and ideologies 83
  intrinsic/extrinsic 133–4, 146, 157–8
  and social movements 102
  systems 133
Veblen, Thorstein 19
visual
  complexity 128
  culture 114, 182
  intelligence 114–15 (*see also* intelligence)
  language 115–16
  literacy 114, 119–20
vital materialism 67
vitality 58, 67, 68, 131

Wahl, Daniel 68, 81, 88
ways of seeing 1, 23, 79, 113, 116, 118, 120, 130–1
wetiko 70
white privilege 110
Whitehead, Alfred North 57
wicked problems 22, 25, 98
Willis, Anne-Marie 25
Wilson, Edward O. 103, 106, 139
work that reconnects, the 140
worldview 1, 17, 25, 28, 51, 55–7, 62, 65, 74, 77, 79, 109, 117, 123, 125, 178, 181, 184